Elementary Algebra

Dedicated to my team

astrarka

DIFFERENTIATED LEARNING PLATFORMS
FOR KIDS

ISBN: 1453837612
EAN-13: 9781453837610

First Edition 2010

Contents

Foreword

Algebra is amongst the oldest of branches in Mathematics. Apart from the music of variables, equations and polynomials, series and progressions – Algebra offers a great deal of insight into solving day to day problems. The origins of the subject can be traced to our ancestors and their endeavours to determine a numerical value of an unknown (which was later formalized as a variable), given a set of inputs and assumptions. Several texts in ancient Greece, Egypt and India, enumerate problems in Algebra which in current day curriculum is covered under the concept of unitary method.

Most students have a blend of emotions while dealing with Algebra. Talking about x and y, unknowns and equations, is exciting, and equally confusing. This is not just our premise, but this is based on feedback and direct inputs we have received from the student, teacher and parent community.

We started a project to address this conceptual gap. We started stringing together a set of concepts followed by a problem set using the preceding texts. This effort left us with a skeletal structure of concepts in Algebra which progresses from basics to through to advanced topics. This formed the basis of our work on "Elementary Algebra".

We sincerely hope that the student is able to get a good grasp of the subject and the techniques after working with the contents of this book.

Astrarka Educational Solutions Private Limited.
Bangalore, India

Preface

We learnt an important lesson from our "Basics of Speed Mathematics" video effort. For this book, we pro-actively worked on the video and paperback simultaneously. We believe that the nature of treatment of the subject, structure and presentation of content is unique. It lends itself to easy reading and a journey through a maze of concepts establishing opportunities for small victories along the way.

The feedback from the parents and teachers, who championed our video products, laid the foundation stone for this book. It has been a humbling experience, although, we must add – that discovering the child in us and going through the materials of elementary school Mathematics filled our work days with immense joy.

It would be impossible for us to acknowledge all the people that have contributed to this mammoth effort. Without their colossal effort, this book would have been an exercise in futility. This is the problem related to completeness in our enumeration. However, it would be unfair if we did not thank a few people whose contributions stood out during the design, production and review process of the video [1] and the book. R Balasubramanian spent long hours in getting the type setting right. Without his tireless efforts to ensure the accuracy of the content, we would not have been able to complete this project.

We would like to thank all the staff of Astrarka for their contributions, support and assistance throughout this project.

August 2010
Chandramouli Mahadevan,
Bangalore.

---

[1] The 10 hour video on Elementary Algebra is a 3 DVD set produced by Astrarka. For additional details about the product and its features, please feel free to contact sales@astrarka.com or http://www.astrarka.com

1 Good Habits

Concept 0:
There are five fundamental principles, or **good habits** that we would like to emphasize before we commence our discussion on Elementary Algebra.

1. Neatness is conducive to accuracy. Refrain from the temptation to write down something quickly and then scratch the same to make the necessary corrections.

2. One of the weaknesses we find in students while solving word problems is the usage of the = sign. This sign has a specific meaning in the world of mathematics. It cannot be used as a way to begin every new line of step in the problem solving process. Use appropriate mathematical signs and symbols. Never use them to mean something vague. = sign is never a good space filler.

3. Spend a second or two to explain how you arrived at a certain step. Several books and references use a statement, such as "it follows from the above statement". We have oftentimes wondered how the expression or equation below follows from the one above. A good explanation is an excellent demonstration of your understanding of the underlying principles.

4. When you are faced with several conclusions during a problem solving process, it is a good idea to number the statements or equations. In subsequent steps, you can refer to these conclusions by using the label or the assigned equation number.

5. The easiest of problems attracts the silliest of mistakes. If the problem is easy, motivate yourself to get it right. Do not let over-confidence or carelessness take control of the situation.

2 Fundamentals

Concept 1:
Algebra is similar to Arithmetic. In both cases, we manipulate quantities. Arithmetic deals with manipulation of numbers. Algebra is slightly different. In Algebra, we deal with a greater generality. We employ letters or symbols to denote quantities on which we perform the mathematical operations or manipulations.

The term **manipulation** means the same in both the worlds. A set of inputs or information about a problem is given. We will derive the result using the inputs. Thus, from a problem-solving standpoint, both Arithmetic and Algebra are very similar.

Each number represents a unique point in the number line. An algebraic symbol stands for one or more numerical values. In several cases, we can have a symbol that can stand for an infinitely large number of numeric values. It is possible for us to apply the mathematical operations without assigning any specific value to a symbol. This may confuse the beginner. This will become clearer as we make progress.

The meanings of the mathematical operations such as addition, subtraction, multiplication and division remain the same in both worlds — Arithmetic and Algebra.

Example: Let us look at the following statement.

Jack is 5 years older than Jill.

Let Jill's age be x years. Then, Jack's age will be $(x + 5)$ years.

In the above formulation, the unknown quantity x is a symbol used to represent the age of Jill. It could be any number of years. We know that Jack's age is simply "Jill's age plus 5 years". This reality is represented as $(x + 5)$ years.

Here x is a variable. It can assume any numerical value as determined by the final solution to the problem. x is also a symbol which stands for Jill's age. This fundamental concept is all-pervasive in Algebra.

Concept 2:
An algebraic term is simply a variable with or without a coefficient.

3

Therefore, x is a variable, a literal or a symbol (we will use these terms interchangeably).

$7x, 8x$ are examples of algebraic terms.

When the coefficient is absent, we assume that it is equal to one.

Therefore, the term x is the same as $1x$.

An algebraic expression is simply a collection of algebraic terms connected by mathematical operations such as $+, -, \times$ and $/$. There are other mathematical operations as well. We will consider those in the subsequent sections, when appropriate.

For example, $7a + 6b + 8c - d - 16z$ is an algebraic expression that has 5 terms in it.

An algebraic term $7x$ is the same as $+7x$. In other words, we assume the sign to be +, whenever the sign is not explicitly specified. This is similar to the arithmetic concept of numbers.

8 is the same as $+8$. And, $197 = +197$.

A simple expression consists of one term.

$8x, 9y$ and $16a$ are examples of a simple expression.

A compound expression consists of several terms.

$8x + 9y - 16a$ is a compound expression.

A binomial is a compound expression with two terms.

A trinomial is a compound expression with three terms.

In addition, a multinomial is a compound expression with more than three terms.

For example, $3x + 2y$ is a binomial expression, $6a + 7b - 3c$ is a trinomial and $8x - 4y - 23z + 16a - 2d$ is a multinomial.

Concept 3:
Just like in arithmetic, when two or more variables, terms or expressions are multiplied together, we get a product. However, there is one big difference in representation.

The product of two variables, a and b, is ab. This is the same as $a \times b$.

If $a = 9$ and $b = 4$, $ab = a \times b = 9 \times 4 = 36$.

In arithmetic, 4×5 is 20 and not 45. We cannot drop the product sign in arithmetic. In fact, 45 is $4 \times 10 + 5$.

Concept 4:
In arithmetic, $24 = 2 \times 3 \times 4$. Therefore, two, three and four are factors of 24. Similarly, in an algebraic term $7abc$ has four factors. These are seven, a, b and c.
It would be important for us to recall that $7abc = 7 \times a \times b \times c$.

Concept 5:
The numeric part of an algebraic term is usually called the **coefficient** of the term. Therefore, 6 is the coefficient of ab in $6ab$. In the algebraic term abc, we can refer to a as the **literal coefficient** of bc.

Therefore, there are situations where a coefficient is not necessarily a numerical quantity. In general, when we simply talk about coefficients, we refer to the numerical part of a term.

The term a is the same as $1a$. So, when the coefficient of a term is 1, we usually ignore this and simply write the variables involved.
Therefore, $abc = 1abc$. Or, $1xyz = xyz$.

Concept 6:
If we consider a quantity or a variable x, and repeatedly multiply the variable by itself, we get a product. This product is called power of the variable or quantity.

Therefore, $a \times a$ is written as a^2. This is the second power of a. This is also read out as "a to the power of 2". This is also called a-squared.

Similarly, $a \times a \times a = a^3$. This is the third power of a. This is is read out as "a to the power of 3". This is also called a-cubed.

And, $a \times a \times a \times a \ldots (n \text{ times}) = a^n$. This is the nth power of a. This is read out as "a to the power of n".

In these examples, a is called the base and the number which represents the power is called the index or exponent.

The quantity a is the same as a^1, which is the first power of a.

Therefore, $a = 1 \times a = a \times 1 = 1 \times a^1$.
They all have the same meaning and the 1 is omitted in each of these cases.

By definition, any base to the power of zero is 1. Zero to the power of any anything is zero.
Therefore, $a^0 = 1$ and $0^n = 0$.

Every power of 1 is equal to 1.
Therefore $1^2 = 1^3 = 1^9 = 1^n$.

Concept 7:
Conceptually, a coefficient and an exponent are very different things. It is critical to spend a moment to understand the difference – since both seem to fit the definition of being the numerical part of a term.

$4a$ means $4 \times a$. $3b$ means $3 \times b$.

a^4 means $a \times a \times a \times a$ or a multiplied by itself four times. Similarly, b^3 means $b \times b \times b$.

If $a = 3$ and $b = 4$, then $4a = 4 \times 3 = 12$; $a^4 = 3 \times 3 \times 3 \times 3 = 81$.
Similarly, $3b = 3 \times 4 = 12$; and $b^3 = 4 \times 4 \times 4 = 64$.

Concept 8:
Like in the arithmetic world, multiplication is commutative in the algebraic world as well.

In other words, just like $4 \times 5 = 5 \times 4$ in the arithmetic world; $ab = ba$ in the algebraic world.

The order in which we multiply the constituent quantities in a term has no impact on the final product.

xyz, yxz and yzx are all equal to one another – they produce the same final product.

Concept 9:
$aaaabbbcccccc$ is the same as $a^4 b^3 c^5$. Similarly, $16x^2 y^3 z^6$ is the same as $16xxyyyzzzzzz$.

In other words, you can multiply several quantities, each raised to some power; the basic notation of multiplication applies with no exception.

Therefore, if $a = 3; b = 4; c = 5$; then $16a^2bc^3 = 16 \times 3 \times 3 \times 4 \times 5 \times 5 \times 5 = 72000$.

Concept 10:
If one of the factors of a product is zero, then the product is always zero, irrespective of the number of other non-zero factors in the product. This can be simply stated as "anything multiplied by zero gives a zero".

Therefore, we can also conclude that every power of zero is zero. $0 = 0^1 = 0^9 = 0^n$.

Concept 11:
There is a distinction between sum of two terms and product of two terms. The sum of two terms a and b is $a + b$; whereas the product of the two terms is ab. We cannot ignore any other mathematical symbol. The only symbol that is ignored in algebra is the multiplication symbol.

Question 2.1:   if $a = 5, b = 4, c = 1, x = 3, y = 12, z = 2$, find the values of

1. $2a$

2. a^2

3. $3z$

4. z^3

5. c^4

6. $4c$

7. $4b^2$

8. c^3

9. x^3

10. $3x$

11. $7y^2$

12. $8a^3$

13. $6z^5$

14. $5z^6$

15. $7c^6$

Question 2.2: if $a = 6, p = 4, q = 7, r = 5, x = 1$, find the value of

1. ap

2. $3pq$

3. $3qx$

4. $5p^3$

5. $8aqx$

6. pqr

7. $8aqr$

8. $7qrx$

9. $2apx$

10. $7x^4$

11. $3p^4$

12. $8r^4$

13. $9apqx$

14. $6x^7$

15. x^{10}

Question 2.3: if $a = 4, b = 1, c = 3, f = 5, g = 7, h = 0$, find the value of

1. $3f + 5h - 7b$

2. $7c - 9h + 2a$

3. $4g - 5c - 9b$

4. $3g - 4h + 7c$

5. $3f - 2g - b$

6. $9b - 3c + 4h$

7. $3a - 9b + c$

8. $2f - 3g + 5a$

9. $3c - 4a + 7b$

10. $3f + 5h - 2c - 4b + a$

11. $6h - 7b - 5a - 7f + 9g$

12. $7c + 5b - 4a + 8h + 3g$

13. $9b + a - 3g + 4f + 7h$

14. $fg + gh - ab$

15. $gb - 3hc + fb$

16. $fh + hb - 3hc$

17. $f^2 - 3a^2 + 2c^3$

18. $b^3 - 2h^3 + 3a^2$

19. $3b^2 - 2b^3 + 4h^2 - 2h^4$

3   Addition, Subtraction and Simple Brackets

Concept 12:
In an arithmetic expression, quantities in an expression prefixed with a + sign are additive terms and those prefixed with a − sign are subtractive terms.

In an algebraic expression, the same concept of additive and subtractive terms exists.

Therefore, in the expression $3a - 2b - 3c$; $3a$ is an additive term and $-2b$ and $-3c$ are subtractive terms.

Concept 13:
The sum of all additive terms in an arithmetic expression is always greater than the sum of all subtractive terms.
In an arithmetic expression $4+7-5-9$; the sum of $+4$ and $+7$ is always greater than the sum of -5 and -9.

The same statement cannot be made in the case of an algebraic expression. It is perfectly possible for the sum of subtractive terms to be larger than the additive terms.
In the above expression $3a - 2b - 3c$, it is possible for $-2b + -3c$ to be greater than $3a$. In fact a subtractive term may also have a meaning even when it stands by itself.

Concept 14:
If you started off with $100; made a profit of $70 and then a loss of $30; then you will be left with $100 + $70 − $30 = $140. In other words, you would have made a profit of $40.

The algebraic statement for this is $140 − $100 = $40

If you started off with $100; made a loss of $70 and then a profit of $30; then you will be left with $100 − $70 + $30 = $60. In other words, you would have made a loss of $40.

The algebraic statement for this is $60 − $100 = −$40

The negative sign indicates net loss.

11

Concept 15:
A subtractive quantity is always the opposite of an additive quantity. This is true of arithmetic and algebra. $+4$ and -4 are equidistant from 0, but on either sides of 0. In that sense, we can look at subtraction as the reverse operation, or opposite of addition.

Concept 16:
Sets of algebraic terms that differ only in the coefficients are called like terms. All terms that are not like terms are called unlike terms.

$3a$, $-7a$ and $12a$ are like terms.
$3a$, $3a^2$ and $-3ab$ are unlike terms.

Concept 17:
Addition of algebraic expression follows the following rules:

1. Collect all like terms together

2. Add all the numerical values of coefficients of all the additive like terms of each set of like terms. Prefix the resultant sum with a + sign.

3. Add all numerical values of coefficients of all the subtractive like terms of each set of like terms. Prefix the resultant sum with a – sign.

4. Add the two resultant sums of the coefficients to get the final sum of each of the sets of like terms.

For example, while adding $12a-26a+13a+6a-2a$, the sum of additive like terms is $12 + 13 + 6 = 31$.
Prefix this with a + sign. We get $+31$.

The sum of subtractive like terms is $26 + 2 = 28$. Prefix this with a - sign. We get -28.

Add these two together. We get $31a - 28a = 3a$.

Concept 18:
The sum of two like terms with the same numerical coefficient but different in sign is always zero.

Therefore $3a - 3a = 0$; $5a - 5a = 0$ and $16xy - 16xy = 0$.

Concept 19:
The algebraic sum is independent of the order in which the terms are added. In other words, $x + y = y + x$, and $x - y = -y + x$.
Similarly, $x + y - z = -z + y + x = y - z + x$.

Concept 20:
We use brackets to group sets of terms together. It is also useful to unambiguously indicate what we are planning to do.
When we say $x + (y + z)$ indicates that we wish to add y and z. Then add the resultant sum to x.

$x + (y - z)$ indicates that we want to add the difference of y and z, to x.
$x + (y - z) = x + y - z$

Concept 21:
When we come across a + sign in front of the brackets; we can safely drop the brackets, with no change to signs of any of the terms in the expression. $x + y - z + (a + b - c - d) = x + y - z + a + b - c - d$

Concept 22:
When we come across a – sign in front of the brackets; we can drop the brackets, and change the sign of each of the terms inside the brackets.
$x + y - z - (a + b - c - d) = x + y - z - a - b + c + d$

Concept 23:
From the above two concepts, the following observations can be made.

1. $a + (+b) = a + b$

2. $a + (-b) = a - b$

3. $a - (-b) = a + b$

4. $a - (+b) = a - b$

Question 3.1:   Find the sum of

1. $2a + 3a + 6a + a + 4a$

2. $4x + x + 5x + 6x + 8x$

3. $6b + 11b + 8b + 9b + 5b$

4. $6c + 7c + 3c + 16c + 18c + 101c$

5. $2p + p + 4p + 7p + 6p + 12p$

6. $d + 9d + 3d + 7d + 4d + 6d + 10d$

7. $-2x - 6x - 10x - 8x$

8. $-3b - 13b - 19b - 5b$

9. $-y - 4y - 2y - 6y - 4y$

10. $-17c - 34c - 9c - 6c$

11. $-21y - 5y - 3y - 18y$

12. $-4m - 13m - 17m - 59m$

13. $3x - 10x - 7x + 12x + 2x$

14. $8ab - 6ab + 5ab - 4ab$

15. $2xy - 4xy - 3xy + xy + 7xy$

Question 3.2:   Find the value of

1. $-9a^2 + 11a^2 + 3a^2 - 4a^2$

2. $b^3 - 2b^3 + 7b^3 - 9b^3$

3. $-11a^3 + 3a^3 - 8a^3 - 7a^3 + 2a^3$

4. $a^2b^2 - 7a^2b^2 + 8a^2b^2 + 9a^2b^2$

5. $a^2x - 11a^2x + 3a^2x - 2a^2x$

6. $2p^3q^3 - 31p^3q^3 + 17p^3q^3$

7. $7m^4n - 15m^4n + 3m^4n$

8. $9abcd - 11abcd - 41abcd$

9. $13pqx - 5pqx - 19pqx$

10. $2x^3 - 3x^3 - 6x^3 - 9x^3$

Question 3.3:   Find the sum of:

1. $(3a + 2b - 5c); (-4a + b - 7c); (4a - 3b + 6c)$

2. $(3x + 2y + 6z); (x - 3y - 3z); (2x + y - 3z)$

3. $4p + 3q + 5r; -2p + 3q - 8r; p - q + r$

4. $7a - 5b + 3c; 11a + 2b - c; 16a + 5b - 2c$

5. $8l - 2m + 5n; -6l + 7m + 4n; -l - 4m - 8n$

6. $5a - 7b + 3c - 4d; 6b - 5c + 3d; b + 2c - d$

7. $2a + 4b - 5x; 2b - 5x; -3a + 2y; -6b + 8x + y$

8. $7x - 5y - 7z; 4x + y; 5z; 5x - 3y + 2z$

9. $5 - x - y; 7 + 2x; 3y - 2z; -4 + x - 2y$

10. $25a - 15b + c; 4c - 10b + 13a; a - c + 20b$

11. $2a - 3b - 2c + 2x; 5x + 3b - 7c; 9c - 6x - 2a$

12. $3a - 5c + 2b - 2d; b + 2d - a; 5c + 3f + 3e - 2a - 3b$

13. $p - q + 7r; 6q + r - p; q - 3p - r; 6q - 7p$

14. $17ab - 13kl - 5xy; 7xy; 12kl - 5ab; 3xy - 4kl - ab$

15. $2ax - 3by - 2cz; 2by - ax + 7cz; ax - 4cz + 7by; cz - 6by$

Question 3.4:   Find the sum of the following expressions:

1. $x^2 + 3xy - 3y^2; -3x^2 + xy + 2y^2; 2x^2 - 3xy + y^2$

2. $2x^2 - 2x + 3; -2x^2 + 5x + 4; x^2 - 2x - 6$

3. $5x^3 - x^2 + x - 1; 2x^2 + 5x + 4; x^2 - 2x - 6$

4. $a^3 - a^2b + 5ab^2 + b^3; -a^3 - 10ab^2 + b^3; 2a^b + 5ab^2 - b^3$

5. $3x^3 - 9x^2 - 11x + 7; 2x^3 - 5x^2 + 2; 5x^3 + 15x^2 - 7x; 8x - 9$

6. $4m^3 + 2m^2 - 5m + 7; 3m^3 + 6m^2 - 2; -5m^2 + 3m; 2m - 6$

7. $ax^3 - 4bx^2 + cx; 3bx^2 - 2cx - d; bx^2 + 2d; 2ax^3 + d$

8. $py^2 - 9qy + 7r; -2py^2 + 3qy - 6r; 7qy - 4r; 3py^2$

9. $5y^3 + 20y^2 + 3y - 1; -2y + 5 - 7y^2; -3y^2 - 4 + 2y^3 - y$

10. $2 - a + 8a^2 - a^3; 2a^3 - 3a^2 + 2a - 2; -3a + 7a^3 - 5a^2$

11. $1 + 2y - 3y^2 - 5y^3; -1 + 2y^2 - y; 5y^3 + 3y^2 + 4$

12. $c^7 - 2c^5 + 11c^6; -2c^7 - 2c^6 + 5c^5; 4c^6 - 10c^5; 4c^7 - c^6$

13. $4h^3 - 7 + 3h^4 - 2h; 7h - 3h^3 + 2 - h^4; 2h^4 + 2h^3 - 5$

14. $x^2 + 2xy + 3y^2; 3z2 + 2yz + y^2; x^2 + 3z^2 + 2xz; z2 - 3xy - 3yz; xy + xz + yz - 6z^2 - 4y^2 - 2x^2$

Question 3.5:   Subtract

1. $a + 2b - c$ from $2a + 3b + c$

2. $2a - b + c$ from $3a - 5b - c$

3. $3x + y - z$ from $x - 4y + 3z$

4. $x + 8y + 8z$ from $10x - 7y - 6z$

5. $-m - 3n + p$ from $-2m + n - 3p$

6. $3p - 2q + r$ from $4p - 7q + 3r$

7. $a - 7b - 3c$ from $-4a + 3b + 8c$

8. $-a - b - 9c$ from $-a + b - 9c$

9. $3x - 5y - 7z$ from $2x + 3y - 4z$

10. $-4x - 2y + 11z$ from $-x + 2y - 13z$

11. $-2x - 5y$ from $x + 3y - 2z$

12. $3x - y - 8z$ from $x + 2y$

13. $m - 2n - p$ from $m + 2n$

14. $2p - 3q - r$ from $2q - 4r$

15. $ab - 2cd - ac$ from $-ab - 3cd + 2ac$

16. $3ab + 6cd - 3ac - 5bd$ from $3ab + 5cd - 4ac - 6bd$

17. $-xy + yz - zx$ from $2xy + zx$

18. $-2pq - 3qr + 4rs$ from $qr - 4rs$

19. $-mn + 11np$ from $-11np$

20. $-x^3 + 3x^2 - x$ from $x^3 - 3x^2 + x$

Question 3.6: When $x = 2, y = 3, z = 4$, find the value of sum of $5x^2$, $-3xy$ and z^2. Also find the value of $3z^x + 3x^y$.

Question 3.7: Add together $3ab + bc - ca, -ab + ca, ab - 2bc + 5ca$. From the sum, take away $5ca + bc - ab$.

Question 3.8: Subtract the sum of $x - y + 3z$ and $-2y - 2z$ from the sum of $2x - 5y - 3z$ and $-3x + y + 4z$.

Question 3.9: Simplify a). $3b - 2b^2 - (2b - 3b^2)$. b). $3a - 2b - (2b + a) - (a - 5b)$

Question 3.10: Subtract $8c^2 + 8c - 2$ from $c^3 - 1$.

Question 3.11:   Add together $3a^2 - 7a + 5$ and $2a^3 + 5a - 3$, and diminish the result by $3a^2 + 2$.

Question 3.12:   Subtract $2b^2 - 2$ from $-2b + 6$, and increase the result by $3b - 7$.

Question 3.13:   Find the sum of $3x^2 - 4x + 8$, $2x - 3 - x^2$ and $2x^2 - 2$, and subtract the result from $6x^2 + 3$.

Question 3.14:   What expression must be added to $5a^2 - 3a + 12$ to produce $9a^2 - 7$?

Question 3.15:   Find the sum of $2x$, $-x^3$, $3x^2$, 2, $-5x$, -4, $3x^3$, $-5x^2$, 8; and arrange the result in ascending powers of x.

Question 3.16:   From what expression must the sum of $5a^2 - 2$, $3a + a^2$ and $7 - 2a$ be subtracted to produce $3a2 + a - 5$.

Question 3.17:   When $x = 6$, find the numberical value of the sum of $1 - x + x^2$, $2x^2 - 1$

Question 3.18:   Find the value of $6ax + (2by - cz) - (2ax - 3by + 4cz) - (cz + ax)$, when $a = 0, b = 1, c = 2, x = 8, y = 3, z = 4$.

Question 3.19:   Subtract the sum of $x^3 - 3x^2$, $2x^2 - 7x$, $8x - 2$, $5 - 3x^3$, $2x^3 - 7$ from $x^3 + x^2 + x + 1$.

Question 3.20:   What expression must be taken from the sum of $p^4 - 3p^3$, $2p + 8$, $2p^2$, $2p^3 - 3p^4$, in order to produce $4p^4 - 3$.

Question 3.21:   What is the result when $-3x^3 + 2x^2 - 11x + 5$ is subtracted from zero.

Question 3.22:   By how much does $b + c$ exceed $b - c$?

Question 3.23:   Find the algebraic sum of three times the square of x, twice the cube of x, $-x^3 + x - 2x^2$, and $x^3 - x - x^2 + 1$.

Question 3.24:   Take $p^2 - q^2$ from $3pq - 4q^2$, and add the remainder to the sum of $4pq - p^2 - 3q^2$ and $2p^2 + 6q^2$.

4   Multiplication

Concept 24:
Multiplication is repeated addition. Therefore:
$$xy = x + x + x + \ldots (y \text{ times}) = yx = y + y + y + \ldots (x \text{ times})$$

Concept 25:
The order of multiplication has no impact on the resultant product. This is also known as the associative law of multiplication.
$$xyz = x(yz) = (xy)z = y(xz)$$

Concept 26:
Coming to the concept of repeated multiplication, we can make the following observations:

$$a^n = a \times a \times a \times \ldots (n \text{ times})$$
$$a^m = a \times a \times a \times \ldots (m \text{ times})$$
$$a^n \times a^m = (a \times a \times a \times \ldots (n \text{ times}) \times (a \times a \times a \times \ldots (m \text{ times})$$
$$= (a \times a \times a \times \ldots (m + n \text{ times})$$
$$= a^{m+n}$$

In these expressions, a is called the base of the term and m and n are called the indices.

Concept 27:
Similarly,

$$a^n = a \times a \times a \times \ldots (n \text{ times})$$
$$a^m = a \times a \times a \times \ldots (m \text{ times})$$
$$a^n / a^m = \frac{(a \times a \times a \times \ldots (n \text{ times})}{(a \times a \times a \times \ldots (m \text{ times})}$$
$$= (a \times a \times a \times \ldots (m - n \text{ times})$$
$$= a^{m-n}$$

Concept 28:
As a consequence of the aforementioned concepts, we can come to the following definitions:

1. $a^0 = 1$, by definition

2. $1^n = 1$

3. $0^n = 0$

4. $0^0 = 0$

Concept 29:
Let is now consider the expression $m(x + y)$:

$$m(x + y) = (x + y)m$$
$$= ((x + y) + (x + y) + (x + y) + \ldots (m \text{ times }))$$
$$= ((x + x + x + \ldots (m \text{ times})$$
$$+ (y + y + y + \ldots (m \text{ times})))$$
$$= mx + my$$

Similarly, $m(x - y) = mx - my$

Therefore, $m(x - y - z) = mx - my - mz$

Concept 30:
Consider, $m(a + b) = ma + mb$
Putting $m = (c + d)$, we have
$$(c + d)(a + b) = (a + b)(c + d) = c(a + b) + d(a + b) = ac + bc + ad + bd$$
Similarly, $(a - b)(c - d) = a(c - d) - b(c - d) = ac - ad - (bc - bd)$

Now, we remove the brackets. We have a $-$ sign in front of the bracket. Therefore the sign of each of the term inside the brackets changes.

$$(a - b)(c - d) = ac - ad - bc + bd$$

Concept 31:
The rule of sign can be summarized from the previous concept. Let us recall the findings.

1. $+a \times +c = +ac$

2. $+a \times -d = -ad$

3. $-b \times +c = -bc$

4. $-b \times -d = +bd$

The rule of signs emerges from the above observations.

1. $+ \times + = +$

2. $+ \times - = -$

3. $- \times + = -$

4. $- \times - = +$

Concept 32:
We can now write down the product of compound expressions based on the concepts we have covered so far.

1. $(x + a)(x + b) = x^2 + xb + ax + ab = x^2 + x(a + b) + ab$

2. $(x - a)(x + b) = x^2 + xb - ax - ab = x^2 + x(-a + b) - ab$

3. $(x + a)(x - b) = x^2 - xb + ax - ab = x^2 + x(a - b) - ab$

4. $(x - a)(x - b) = x^2 - xb - ax + ab = x^2 - x(a + b) + ab$

Concept 33:
The conclusions drawn in the previous concept lead us to the technique for writing down the products based on inspection.

When the compound expression is of the form $(x + a)(x + b)$, we can note the following:

1. The product has three terms.

2. The x^2 term has a coefficient of 1

3. The x term has a coefficient of $(a + b)$. We need to ensure that we take the signs of a and b into account as well.

4. The constant term is simply the product of a and b. Again, we need to take the signs of a and b into account.

Question 4.1:   Find the value of:

1. $5x \times 7$

2. $2b \times 3$

3. $x^2 \times x^3$

4. $5x \times 6x^2$

5. $6c^3 \times 7c^4$

6. $9y^2 \times 5y^5$

7. $3m^2 \times 5m^5$

8. $4a^4 \times 6a^6$

9. $3x \times 4y$

10. $5a \times 6b^2$

11. $4c^2 \times 5d^5$

12. $3p^4 \times 5q^5$

13. $6ax \times 5ax$

14. $3qr \times 4qr$

15. $ab \times ab$

16. $3ac \times 5ad$

17. $a^3x \times a^4x^3$

18. $3x^3y^2 \times 4y^5$

19. $a^3b^5 \times a^5b^4$

20. $a^4 \times 3a^5b^2$

Question 4.2: Multiply

1. $ab - ac$ by a^2c
2. $x^2y - x^2z + 4yz^5$ by x^3yz^3
3. $5a^2 - 3b^2$ by $3ab^2c^4$
4. $a^2b - 5ab + 6a$ by $3a^3b$
5. $a^2 - 2b^3$ by $3x^2$
6. $2ax^2 - b^2y + 3$ by a^2xy
7. $7p^2q - pq^2 + 1$ by $2p^2$
8. $m^2 + 5mn - 3n^2$ by $4m^2n$
9. $xy^2 - 3x^2z - 2$ by $3yz$
10. $a^3 - 3a^2x$ by $2a^2bx$

Question 4.3: Multiply together

1. $a, -2$
2. $-3, 4x$
3. $-x^2, -x^3$
4. $-5m, 3m^3$
5. $-4q, 3q^2$
6. $-4y^2, -4y^3$
7. $-3m^3, 3m^3$
8. $4x^4, -4x^4$
9. $-3x, -4y$

10. $-5a^2, 4x$

11. $-3p^2, -4q^5$

12. $3ab, -4ab$

13. $3a^2, -b^2, 2ab$

14. $-a, -b, -c^2$

15. $3a^2, -2b, -4c^3, -d$

16. $6a^2 - 5a^2b - 4ab^2, -3ab^2$

17. $-ab + ac - bc, -ab$

18. $a^2c - ac^2 + c^4, -a^3c$

19. $-3a^2 - 4ax + 5x^2, -a^2x^3$

20. $-2ab + cd - ef, -3x^2y^2$

Question 4.4: Find the product of:

1. $a + 7, a + 5$

2. $x - 3, x + 4$

3. $a - 6, a - 7$

4. $y - 4, y + 4$

5. $x + 9, x - 8$

6. $c - 8, c + 8$

7. $k + 5, k - 5$

8. $m - 9, m + 12$

9. $x - 12, x + 11$

10. $a - 14, a + 1$

11. $p - 10, p + 10$

12. $d + 7, d + 7$

13. $x - 4, -x + 4$

14. $-y + 3, -y - 3$

15. $-a + 4, -a + 5$

16. $3y - 5, y + 7$

17. $5m - 4, 7m - 3$

18. $7b + c, 7b - 2c$

19. $xy + 2b, xy - 2b$

20. $3x - 4y, 2a + 3b$

Question 4.5:   Multiply together

1. $x^2 - 3x - 2$ by $2x - 1$

2. $4a^2 - a - 2$ by $2a + 3$

3. $2y^2 - 3y + 1$ by $3y - 1$

4. $3x^2 + 4x + 5$ by $4x - 5$

5. $2a^2 - 3a - 6$ by $a - 2$

6. $5b^2 - 2b + 3$ by $-2b - 3$

7. $3x^2 - 2x + 7$ by $2x - 7$

8. $5c^2 - 4c + 3$ by $-2c + 1$

9. $x^2 + x - 2$ by $x^2 + x - 2$

10. $x^2 + x - 2$ by $x^2 - x + 2$

11. $a + 3$ by $a - 2$

12. $a - 7$ by $a - 6$

13. $x - 4$ by $x + 5$

14. $b - 6$ by $b + 4$

15. $y - 7$ by $y - 1$

16. $a - 1$ by $a - 9$

17. $c - 5$ by $c + 4$

18. $x - 9$ by $x - 3$

19. $y - 4$ by $y + 7$

20. $a - 3$ by $a + 9$

5 Division

Concept 34:
Division is the only operation that produces two results. While the primary purpose of division is to determine the quotient, the operation also leaves behind a remainder; which is sometimes positive non-zero; but always less than the divisor. This means the remainder can never be negative or a number or an expression greater than the divisor.

Concept 35:
This can be represented as: dividend / divisor = quotient; remainder

Therefore: quotient × divisor + remainder = dividend

When remainder is equal to zero, this expression reduces to: quotient × divisor = dividend

Concept 36:
Each multiplication fact yields two division facts. Therefore, the rule of sign holds for division.

1. $+a/+b = +c$

2. $+a/-b = -c$

3. $-a/+b = -c$

4. $-a/-b = +c$

Just like the multiplication, clearly, like signs produce + and unlike signs produce –.

Concept 37:
We have already seen $a^m/a^n = a^{(m-n)}$

Concept 38:
Dividing each term of the expression by the divisor term completes the

division of an expression by a term. We retain or change the signs of the terms in the expression based on the rule of signs.

Therefore, $\dfrac{ax^2 + bx + c}{m} = \dfrac{ax^2}{m} + \dfrac{bx}{m} + \dfrac{c}{m}$, where m may be any algebraic term.

Concept 39:
We now address the method for dividing one algebraic expression by another algebraic expression.

1. Step 1: Arrange the dividend and divisor in descending powers of some literal in the expression. In most expressions, this literal is x; but this is not the rule. We will assume that this is x for purposes of explaining this rule. We also insert missing powers of x in the dividend and divisor with a 0 coefficient.

2. Step 2: Divide the term with the highest power of x in the dividend, with the term with the highest power of x in the divisor. We get the partial quotient.

3. Step 3: Multiply the partial quotient with the divisor and subtract this expression from the dividend.

4. Step 4: We then bring down the necessary terms of the dividend.

5. Step 5: We repeat step 2 to step 5 until all terms of dividends are brought down.

Question 5.1: Divide:

1. $2x^3$ by x^3

2. $6a^5$ by $3a$

3. $5a^7$ by a^4

4. $21b^7$ by $7b^3$

5. x^3y^2 by $-xy$

6. $-3xy^3$ by $3y$

7. $4p^2q^3$ by $-2pq$

8. $15m^3n$ by $-5mn$

9. $-l^3m^2$ by $-lm$

10. $-48x^9$ by $-6x^3$

11. $35z^{11}$ by $-7z^5$

12. $-7a^3b$ by $-7b$

13. $-28p^5q$ by $28p^5$

14. $24xyz^3$ by $-3z^2$

15. $-12b^2c^5$ by $6b^2c^5$

16. $-9k^{11}$ by $-k^{11}$

17. $2k^3l^5$ by $-kl$

18. $-45a^4b^3c^{15}$ by $9a^2b^3c^{10}$

19. $-186a^2b^2cx^2$ by $-7abx^2$

20. $5a^3b - 7ab^3$ by ab

21. $3x^2 - 2x$ by x

22. $x^2 - xy - xz$ by $-x$

23. $10a^3 - 5a^2b + a$ by $-a$

24. $4x^3 + 36ax^2 - 16x$ by $-4x$

25. $3a^3 - 9a^2b - 6ab^2$ by $-3a$

Question 5.2:   Divide:

1. $a^2 + 2a + 1$ by $a + 1$

2. $b^2 + 3b + 2$ by $b + 2$

3. $x^2 + 4x + 3$ by $x + 1$

4. $y^2 + 5y + 6$ by $y + 3$

5. $x^2 + 5x - 6$ by $x - 1$

6. $x^2 + 2x - 8$ by $x - 2$

7. $p^2 + 3p - 40$ by $p + 8$

8. $q^2 - 4q - 32$ by $q + 4$

9. $a^2 + 5a - 50$ by $a + 10$

10. $m^2 + 7m - 78$ by $m - 6$

11. $x^2 + ax - 30a^2$ by $x + 6a$

12. $a^2 + 9ab - 36b2$ by $a + 12b$

13. $2x^2 - 13x - 24$ by $2x + 3$

14. $5x^2 + 16x + 3$ by $x + 3$

15. $6x^2 + 5x - 21$ by $2x - 3$

16. $12a^2 + ax - 6x^2$ by $3a - 2x$

17. $-5x^2 + xy + 6y^2$ by $-x - y$

18. $6a^2 - ac - 35c^2$ by $2a - 5c$

19. $12p^2 - 74pq + 12q^2$ by $2p - 12q$

20. $12^2 - 31ab + 20b^2$ by $4a - 5b$

6 Elementary Rules

Concept 40:
We know that $a \times a = a^2$. a is called the **square root** of a^2.
This is also denoted as $a = \sqrt{a^2}$.

Concept 41:
Similarly, a is the cube root of a^3; a is the fourth root of a^4 and a is the fifth root of a^5.

Generalizing this observation, we have a is the nth root of a^n.

Concept 42:
The number of literals in an algebraic term, excluding the numerical coefficient, is the degree of the term. Each literal constitutes a dimension of the term.

Therefore, $4xyz$ has three dimensions and is of third degree.

Concept 43:
The highest dimensions in an expression are also known as the degree of an expression.

For example, $ax^2 + bx + c$ is an expression in second degree in x.

Concept 44:
When all terms in a compound expression are of the same dimensions, we call the expression a homogenous expression.

For example, $a^3 + b^3 + 3a^2b + 3ab^2$ is a homogenous compound expression.

The product of two or more homogenous expression is always homogenous.

Similarly, when we divide a homogenous expression by another homogenous expression, the quotient is also homogenous.

Concept 45:
This concludes our review of the concepts of basic operations like addition, subtraction, multiplication and division of algebraic terms and expressions. We have used arithmetic operations and principles as the golden reference. We have compared the corresponding algebraic realities against this gold standard.

Question 6.1: Simplify by removing brackets

1. $3(x - 2y) - 2(x - 4y)$

2. $x - 3(y - x) - 4(x - 2y)$

3. $16 - 3(2x - 3) - (2x + 3)$

4. $4(x + 3) - 2(7 + x) + 2$

5. $8(x - 3) - (6 - 2x) - 2(x + 2) + 5(5 - x)$

6. $2x - 5(3x - 7 + y) + 4(2x + 3y - 8) - 7y$

7. $2x - 5(3x - 7(4x - 9))$

8. $x^3 + 3(x^2y + xy^2) + y^3 - x^3 - 3(x^2y - xy^2) - y^3$

9. $4x - 3(x - (1 - y) + 2(1 - x))$

10. $x - (y - z)(x - y - z - 2(y + z))$

Question 6.2: Find the sum of $a - 2b + c, 3b - (a - c), (3a - b) + 3c$.

Question 6.3: Subtract $1 - x^2$ from 1, and add the result to $2y - x^2$

Question 6.4: Simplify $a + 2b - 3c + (b - 3a + 2c) - (3b - 2a - 2c)$.

Question 6.5: Find the product of $3x^2y, 2xy^2, -7x^3, -5x^4y^5$.

Question 6.6: Simplify $2x^2 - (2xy - 3y^2) + 4y^2 + (5xy - 2x^2) + x^2 - (2xy + 6y^2)$.

7 Simple Equations

Concept 46:
We now turn our attention to the important concept of an equation. This is a central notion in algebra.

When we assign the resultant value of an algebraic term or an expression to another algebraic term or an expression or a number; we refer to this assignment of equality as an equation.

Therefore, we can enumerate a few examples of equations using this definition.

1. $2x + 5 = 9y$

2. $7y + 2x = 5$

Concept 47:
Each equation has three parts. An "equal" sign that separates the expression to its left and the expression to its right constitutes an equation.

The most generic form of an equation therefore is:
LHS expression = RHS expression

Concept 48:
An equation that is valid for all values of x, the unknown is called an identity.

For example:
$(x+a)^2 = x^2 + 2ax + a^2$ is an identity because this is valid for all values of x and a.
x and a are the unknowns in this equation.

Concept 49:
Certain equations are conditionally true. This means that these equations are true only for certain values of x or their unknowns. These are called conditional equations.

1. $9x = 27$ is valid only if $x = 3$ and for no other value of x.

2. $16x = 256$ is valid for $x = 16$ and for no other value of x.

3. $x^2 = 81$ is true when $x = 9$ or $x = -9$ and for no other value of x.

In these examples, the solutions simply are values of x that satisfy the equation.

In $3x = 9$; $x = 3$ is a solution.
3 is said to satisfy the equation.
x is the unknown quantity in the equation.

The process of determining the value of the unknown quantity that satisfies the equation is called solving the equation.

Concept 50:
An equation that involves an unknown quantity in the first degree is also known as simple equation. The solution to a simple equation can be determined by applying the following axioms to the equations:

1. Adding the same quantity to both sides of the equation does not affect the equality.

2. Subtracting the same quantity to both sides of the equation does not affect the equality.

3. Multiplying the same quantity to both sides of the equation does not affect the equality.

4. Dividing the same non-zero quantity to both sides of the equation does not affect the equality.

Concept 51:
Principle of transposition follows from the axioms mentioned in the preceding section. We will derive these from first principles.

Consider the equation: $x = a$
Subtract a from both sides of the equation; the equality is not impacted.

$x - a = a - a$

$x - a = 0$

This means that a can be transposed from RHS to the LHS with a change in sign. Generalizing the observation, we can conclude that any term can be transposed from one side of the equation to the other side by changing its sign.

Consider the equation: $mx = a$

Dividing both sides by m makes no change in the equality.

$\dfrac{mx}{m} = \dfrac{a}{m}$

$x = \dfrac{a}{m}$

Thus a term which is a factor on one side, becomes the divisor on the other side. Similarly a divisor on one side of the equation becomes a factor on the other side after a transposition operation.

Concept 52:
The process for solving simple equations is straightforward.

1. We transpose all terms involving the unknown quantity to the left hand side of the equation.

2. We transpose all other terms to the right hand side of the equation.

3. We divide both sides of the equation by the coefficient of the unknown quantity to get the value of the unknown quantity.

Concept 53:
Verification of the solution is an important part of the problem solving process. Verification ensures accuracy of the solution. It is also recommended that you substitute the solution in the original equation and ensure that the equation is satisfied.

Question 7.1: Write down the solution to the following equations:

1. $7x = 21$
2. $3x = 15$
3. $9x = 18$
4. $5x = 5$
5. $12x = 132$
6. $33 = 11x$
7. $4x = -12$
8. $-10 = -5x$
9. $4x = 18$
10. $12x = 42$
11. $30 = -6x$
12. $4x = 0$
13. $6x = 26$
14. $0 = 11x$
15. $1 = 11x$
16. $3x = -27$
17. $0 = -2x$
18. $6x = 3$
19. $5 = 15x$
20. $-24 = -8x$

Question 7.2: Solve the following equations:

1. $6x + 3 = 15$

2. $5x - 7 = 28$

3. $13 = 7 + 2x$

4. $15 = 37 - 11x$

5. $4x - 7 = 11$

6. $7x = 18 - 2x$

7. $3x - 18 = 7 - 2x$

8. $4x = 13 - 2x - 10$

9. $3x = 7 - 2x + 8$

10. $5x - 17 + 3x - 5 = 6x - 7 - 8x + 115$

11. $5(x - 3) = 4(x - 2)$

12. $11(5 - 4x) = 7(5 - 6x)$

13. $3 - 7(x + 1) = 5 - 4x$

14. $5 - 4(x - 3) = x - 2(x - 1)$

15. $8(x - 3) - 2(3 - x) = 2(x + 2) - 5(5 - x)$

16. $(x + 2)(x + 3) + (x - 3)(x - 2) - 2x(x + 1) = 0$

17. $(2x + 1)(2x + 6) - 7(x - 2) = 4(x + 1)(x - 1) - 9x$

18. $(3x + 1)^2 + 6 + 18(x + 1)^2 = 9x(3x - 2) + 65$

19. Show that $x = 5$ satisfies $5x - 6(x - 4) = 2(x + 5) + 5(x - 4) - 6$

20. Verify that $x = 3$ satisfies $2(x + 1)(x + 3) + 8 = (2x + 1)(x + 5)$

8 Symbolic Expressions

Concept 54:
Armed with the fundamentals for manipulating algebraic expressions, we can now look at concepts of symbolic expressions. Algebra may be treated as a mathematical language used for representing real world problems that we face. These problems typically appear in English or your native language. Algebra ensures that the problems are represented by means of symbols, expressions and relationships, which can be looked at as a whole and solved.

Express the solutions to the following problems as symbolic expressions.

Question 8.1: By how much does x exceed 5?

Question 8.2: By how much is y less than 15?

Question 8.3: What must be added to a to make 7?

Question 8.4: By what must 5 be multiplied to make a?

Question 8.5: What is the quotient when 3 is divided by a?

Question 8.6: By what must $6x$ be divided to get 2?

Question 8.7: By how much does $6x$ exceed $2x$?

Question 8.8: The sum of two numbers is x and one of the numbers is 10. What is the other number?

Question 8.9: The sum of three numbers is 100 and one of the numbers is 25, other is x, what is the third number?

Question 8.10: The product of two numbers is $4x$. One of the factors is 4. What is the other factor?

Question 8.11:   The product of two number p and one of them is m. What is the other number?

Question 8.12:   How many times is x contained in $2y$?

Question 8.13:   The difference between two numbers is 8, and the greater of them is a; what is the other?

Question 8.14:   The difference between two numbers is x; and the lesser of them is 6; what is the other?

Question 8.15:   What number is 30 less than y?

Question 8.16:   The sum of 12 equal numbers is $48x$. What is the value of each number?

Question 8.17:   How many numbers, each of value y must be taken to make $15xy$?

Question 8.18:   If there are x numbers each equal to $2a$; what is their sum?

Question 8.19:   If there are 5 numbers each equal to x; what is their product?

Question 8.20:   If there are x numbers each equal to p, what is their product?

Question 8.21:   If there are n books each worth Rs. y; what is the total cost of the books?

Question 8.22:   How many books each costing Rs. 2 be bought for Rs. y?

Question 8.23:   What is the cost of x apples at Rs. y a dozen?

Question 8.24:   What is the cost of n oranges at Rs. m a score?

Question 8.25:   How many hours will it take to travel x miles at y miles an hour?

Question 8.26:   How far can I walk in p hours at the rate of q miles per hour?

Question 8.27:   If I can walk m miles in n days, how many miles am I walking every day?

Question 8.28:   In 5 years from now, my age will be half of that of my father. How old is my father now, if my age is x years today?

Question 8.29:   The sum of 3 consecutive numbers is 27. What are the numbers?

Question 8.30:   The number of 2 consecutive odd numbers is always divisible by 4. Is this true?

Question 8.31:   The sum of three consecutive odd numbers is 33. What are the numbers?

Question 8.32:   What is the next odd number after $2n - 1$?

Question 8.33:   How old is a man who in x years will be n times as old as his son's age today, given that his son is y years old today?

Question 8.34:   How old were you x years ago, given that you will be y years old, z years from now?

Question 8.35:   What is the cost of $6x$ plums and $4x$ nuts; given that a nut costs c and m plums are y times the cost of z nuts?

9 HCF, LCM and Elementary Fractions

Concept 55:
The common factor which is of the highest dimension, that divides a set of expressions without leaving a remainder is called the highest common factor. This is also referred to as HCF — highest common factor or GCF — greatest common factor or GCM — greatest common measure.

We will use HCF to denote this.

Concept 56:
In case of simple expressions, we can determine the HCF by inspection.

1. Step 1: Determine the HCF of the numerical coefficients of the expressions or algebraic terms.

2. Step 2: For each literal check the highest power which will divide each of the expression without leaving a remainder. This is also the least power of that literal across all the expressions.

3. Step 3: The HCF is simply the product of step 1 and step 2.

For example, HCF of $6a^3b^6$, $9a^2b^9$ and $12a^6b^7$ is $3a^2b^6$.

Concept 57:
The least common multiple is simply the expression with the lowest dimension, which is divisible by each of the expressions without leaving a remainder. This is abbreviated as LCM.

Concept 58:
We can determine the LCM of simple expressions by inspection.

1. Step 1: Determine the LCM of the numerical coefficients of the expressions or algebraic terms.

2. Step 2: For each literal check the lowest power which is divisible by each of the expression without leaving a remainder. This is also the highest power of that literal across all the expressions.

3. The LCM is simply the product of step 1 and step 2.

For example, LCM of $6a^3b^6$, $9a^2b^9$ and $12a^6b^7$ is $36a^6b^9$.

Concept 59:
The concept of fractions is the same in arithmetic and algebra. Let us commence our discussion of fractions with the definition of fractions. If a whole is split into b equal parts, and we consider a of those parts; we are talking of $\frac{a}{b}$ of the whole.

In arithmetic, the whole is usually treated as 1. The whole in the world of algebra is an unknown quantity x. The resultant fraction is $\frac{a}{b}$ of x. It refers to a parts of b equal parts that x is divided into.

Concept 60:
In order to reduce the fraction to its lowest terms, we eliminate the common factors from the numerator and denominator. The factor that remains behind after this process of canceling common factors is the lowest term of the fractions.

Concept 61:
The product of two algebraic fractions is similar to the case in arithmetic. We divide the product of numerators by the product of denominators. We complete the operation by converting the resultant fraction to lowest terms.

Concept 62:
The reciprocal of a fraction $\frac{a}{b}$ is $\frac{b}{a}$. The notion of a reciprocal is important to the concept of division in fractions.

Concept 63:
To divide one fraction by another fraction, we multiply the first fraction by the reciprocal of the second fraction.

In other words: $\dfrac{\frac{a}{b}}{\frac{c}{d}} = \dfrac{a}{b} \times \dfrac{d}{c}$

Concept 64:
The concept of equivalent fractions is central to the operations of addition and subtraction of fractions. Let us now take a look at the concept of equivalence. This is similar to the world of arithmetic.

1. We take the LCM of the denominators of the fractions.

2. We multiply the numerators by the quotient that we get by dividing the LCM by the denominators.

3. The resultant fractions are equivalent fractions.

Let us consider the following example.

Find equivalent fractions of $\dfrac{a}{3xy}, \dfrac{b}{6xyz}, \dfrac{c}{2yz}$

The LCM of denominators $3xy, 6xyz, 2yz$ is $6xyz$

Therefore $\dfrac{a}{3xy} = a \times \dfrac{2z}{6xyz} = \dfrac{2az}{6xyz}$

Then $\dfrac{b}{6xyz} = \dfrac{b}{6xyz}$ (no change required since denominator = LCM)

And $\dfrac{c}{2yz} = \dfrac{3cx}{6xyz}$

The equivalent fractions of $\dfrac{a}{3xy}, \dfrac{b}{6xyz}, \dfrac{c}{2yz}$ is:

$\dfrac{2az}{6xyz}, \dfrac{b}{6xyz}, \dfrac{2az}{6xyz}$ respectively.

Concept 65:
In order to add or subtract fractions,

1. Step 1: Convert the fractions to its lowest terms.

2. Step 2: Convert the fractions to equivalent fractions.

3. Step 3: Add or subtract the numerators as required

4. Step 4: The common denominator of the resultant fraction is the common denominators of the equivalent fractions.

5. Step 5: Reduce the fraction to lowest terms if required.

Question 9.1: Find the highest common factor of:

1. $3ab^2, 2ab^3$

2. $x^3y^2, 4x^2y^5$

3. $2x^3y^2, 4x^4y^5$

4. $4x^5, 2xy^2z^3$

5. a^2b^2c, a^3bc^3

6. $3a^2b, 9abc$

7. $6x^2y^2z, 2xy$

8. $15y^3, 5xy^6z^2$

9. $12a^3bc^2, 18ab^2c^3$

10. $7x^3y^5z^4, 21x^2yz^3$

Question 9.2: Find the lowest common multiple of:

1. $7x^3y^5z^4, 21x^2yz^3$

2. a^2b^4, abc

3. $2x^3y, 3xy^2z$

4. $4a^2, 3abx^4$

5. $4a^4bc^3, 5ab^2$

6. $2ab, 4xy$

7. mn, nl, lm

8. $xy^2, 3yz^2, 2zx^3$

9. $2xy, 3yz, 4zx$

10. p^2qr, pq^2r, pqr^2

Question 9.3:   Reduce to lowest terms

1. $\dfrac{2a}{4ab}$

2. $\dfrac{3a^2}{9ab}$

3. $\dfrac{2bc^2}{6b^2c}$

4. $\dfrac{2abc}{8a^2bc^2}$

5. $\dfrac{xy^2z^3}{x^3y^4z}$

6. $\dfrac{12mn}{15lm}$

7. $\dfrac{14xy^3}{21x^3z^2}$

8. $\dfrac{9a^3b}{12ab^3c}$

9. $\dfrac{15a^2b^2c^3}{18abc^2}$

10. $\dfrac{5a^3y^2z^4}{15ay^4z}$

Question 9.4:   Simplify the following expressions:

1. $\dfrac{xy}{ab} \times \dfrac{a^2b^3}{xy^2}$

2. $\dfrac{ab}{2cd^3} \times \dfrac{4c^2d}{ab^3}$

3. $\dfrac{2ax^2}{3y^3z} \times \dfrac{yz^3}{4a^2x}$

4. $\dfrac{6a^2x^3}{7ab^2} \times \dfrac{14b^2c}{12ax}$

5. $\dfrac{3ab^2}{5b^3c} \times \dfrac{15b^2c^2}{9a^2b}$

6. $\dfrac{3p^2q^2}{9xy} / \dfrac{pq}{x^2y^2}$

7. $\dfrac{10b^2}{4x^2} / \dfrac{b^2p^2}{3x^6}$

8. $\dfrac{17y}{x^2z^3} / \dfrac{34y^3}{x^5z}$

9. $\dfrac{9ax^2}{5a^2z} / \dfrac{x^3y^2}{2a^2y}$

10. $\dfrac{14d^3}{abc} / \dfrac{81d^3}{27a^2b^2c^2}$

Question 9.5: Simplify the following expressions:

1. $\dfrac{a}{2} + \dfrac{a}{3}$

2. $\dfrac{b}{3} + \dfrac{b}{4}$

3. $\dfrac{x}{4} - \dfrac{x}{5}$

4. $\dfrac{2y}{3} + \dfrac{y}{6}$

5. $\dfrac{a}{5} - \dfrac{a}{6}$

6. $\dfrac{m}{8} - \dfrac{2n}{20}$

7. $\dfrac{2a}{3} + \dfrac{4a}{9b}$

8. $\dfrac{ab}{3} - \dfrac{x^2 y}{6xy}$

9. $\dfrac{a}{xy} + \dfrac{2a}{yz} - \dfrac{3a}{xz}$

10. $\dfrac{a^3}{3a^2 b} - \dfrac{a^3}{ab^2} + \dfrac{ac}{6bc}$

10 Simultaneous Equations

Concept 66:
When the same number of unknown quantities satisfies two or more equations, the set of equations are called **simultaneous equations**. Therefore, two equations in two unknowns, three equations in three unknowns and n equations in n unknowns form a set of simultaneous equations.

We will use the case of 2 equations and 2 unknowns to describe the subsequent concepts; but it must be borne in mind that these are applicable to the case where we have n equations in n unknowns.

Concept 67:
The values of the unknowns that satisfy simultaneous equations form the **solution**. When we solve simultaneous equations, we are asking the key question "What is the value of x and y that will satisfy both the given equations?"

Concept 68:
The process of elimination is the primary technique for solving a simultaneous equation. We try to explore ways of obtaining one equation in one unknown from the simultaneous equations in x and y. There are a couple of ways of doing this.

Concept 69:
Let us consider the following equations:

$$ax + by = m \tag{1}$$
$$cx + dy = n \tag{2}$$

We can transpose the terms in (1), and express one variable in terms of the other. Therefore:

$$x = \frac{m - by}{a} \tag{3}$$

Substituting the value of x in (2), we get the fourth equation which contains only one variable y. Then we can determine the value of y through appropriate transposition of terms and coefficients.

Once we know the value of y; we can substitute the value of y in (3) and determine x. This would lead us to the desired solution to the simultaneous equations.

Remember, we could have expressed y in terms of x as well. As in:

$$y = \frac{m - ax}{b} \tag{4}$$

In this case, we would substitute the value of y in (2). This would result in an equation in one variable—x. Once we determine the value of x, we can determine the value of y through a process of substitution like before.

Concept 70:

The process of elimination can be achieved through an alternative route.

Let us consider the following equations:

$$ax + by = m \tag{5}$$
$$cx + dy = n \tag{6}$$

Multiplying (5) by c and (6) by a, would lead us to two equations where the coefficients of x are the same.

$$acx + bcy = mc \tag{7}$$
$$acx + ady = na \tag{8}$$

Subtracting (7) from (8), we get:

$$y(ad - bc) = an - cm \implies y = \frac{an - cm}{ad - bc} \tag{9}$$

The rest of the steps in the problem solving are similar to the process of elimination we discussed in the previous concept.

Concept 71:

In the previous concept, we accomplished the same effect of eliminating a variable by multiplying the equations with an appropriate factor. How do we determine the factor ?

If we decide to eliminate x from the equations, we determine the LCM of the coefficients of x in the two equations. We multiply equation (1) by $\dfrac{\text{LCM}}{c}$ and equation (2) by $\dfrac{\text{LCM}}{a}$

If we decide to eliminate y from the equations, we would essentially start the process with the LCM of coefficients of y.

Question 10.1: Solve the following equations:

1. $x + y = 19, x - y = 7$

2. $x + y = 23, x - 5 = 5$

3. $x + y = 11, x - y = -9$

4. $x + y = 24, x - y = 0$

5. $x + 6 = 6, x - y = 0$

6. $x - y = 25, x + y = 13$

7. $3x + 5y = 50, 4x + 3y = 41$

8. $x + 5y = 18, 3x + 2y = 41$

9. $4x + y = 10, 5x + 7y = 47$

10. $7x - 6y = 25, 5x + 4y = 51$

11. $11x - 7y = 43, 2x - 3y = 13$

12. $4x - 3y = 0, 7x - 4y = 36$

13. $2x + 3y = 22, 5x + 2y = 0$

14. $7x + 3y = 65, 7x - 8y = 32$

15. $3x - 2y + z = 4, 2x + 3y - z = 3, x + y + z = 8$

16. $3x + 4y - 6z = 16, 4x + y - z = 24, x - 3y - 2z = 1$

17. $x + 2y + 3z = 32, 4x - 5y + 6z = 27, 7x + 8y - 9z = 14$

18. $x - y + z = 5, 6x + 3y + 2z = 84, 3x + 4y - 5z = 13$

11 Involution

Concept 72:
Involution is simply the process of multiplying an expression or term with itself to find the second, third or n^{th} power of the expression or term.

Concept 73:
Let us look at the following set of statements:

$$a^m = a \times a \times a \ldots m \text{ times}$$

$$
\left.
\begin{array}{l}
(a^m)^n = (a \times a \times a \ldots m \text{ times}) \\
\quad \times (a \times a \times a \ldots m \text{ times}) \\
\quad \times (a \times a \times a \ldots m \text{ times}) \\
\quad \ldots
\end{array}
\right\} \ldots n \text{ times}
$$

$$(a^m)^n = (a \times a \times a \ldots m \times n \text{ times})$$

$$= a^{mn}$$

Concept 74:

1. Every even power of an expression is always positive.

2. Every odd power of an expression retains the sign of the original expression.

Concept 75:
To find out the nth power of an expression

1. Raise the numeric coefficient to the required power

2. Prefix the same with the appropriate sign based on the previous Concept

3. Raise the expression to the required power using the previous Concept

4. The product of the previous three steps is the required power of the algebraic expression

---

Concept 76:
A few common expressions

1. $(a + b)^2 = a^2 + 2ab + b^2$

2. $(a - b)^2 = a^2 - 2ab + b^2$

3. $(a + b + c)^2 = a^2 + b^2 + c^2 + 2ab + 2bc + 2ac$

The general approach to finding the square of a multinomial expression is as follows.

1. First create an expression with the sum of squares of each of the terms in the multinomial. Squares are always positive no matter the sign associated with the term is.

2. Now add twice the product of each of the terms with every other term. Ensure that the sign of each term is accounted for while taking the product of two terms.

We can also look at the expression for cubes of sum of terms.

1. $(a + b)^3 = (a + b)(a + b)^2 = (a + b)(a^2 + 2ab + b^2) = a^3 + b^3 + 3a^2b + 3ab^2$

2. $(a - b)^3 = (a - b)(a - b)^2 = (a - b)(a^2 + b^2 - 2ab) = a^3 - b^3 - 3a^2b + 3ab^2$

Question 11.1: Write down the squares of each of the following expressions:

1. $a^2 b$

2. $3ac^3$

3. $5xy^2$

4. $6b^3 c^2$

5. $4a^2 bc^3$

6. $-3x^2 y^5$

7. $-2a^2 b^3 c$

8. $-3dx^4$

9. $a + b$

Question 11.2: Write down the values of each of the following expressions

1. $(ab^2)^4$

2. $(-x^2 y)^5$

3. $(-2m^2 n^3)^6$

4. $(-x^3 y^2)^7$

5. $(-6y^7)^4$

Question 11.3: Write down the cubes of each of the following expressions

1. $2x$

2. $3ab^2$

3. $4x^3$

4. $-3a^2b$

5. $-4x^3y^2$

6. $-b^2cd^3$

7. $-6y^4$

8. $-4p^3q^5$

9. $a^2 - b$

Question 11.4: Write down the values of each of the following expressions

1. Square of $x + 2y$

2. Square of $2ab - xy$

3. Square of $x + y - a - b$

4. Cube of $x + 3y$

5. Cube of $4y^2 - 3$

12 Evolution

Concept 77:
Evolution is the opposite of involution. The operation of finding the nth power of a term is involution. The operation of finding the nth root of a number is evolution.

Evolution of a term x, is another term y such that $x = y^n$. Therefore the nth root of x is y.

Concept 78:
The rule of signs for evolution can be summarized as below:

1. An even root of any positive term or quantity can be either positive or negative. This is because product of two positive terms or two negative terms is always positive.

2. A negative term or quantity cannot have an even root.

3. Every odd root of a term or quantity retains the sign of the term or quantity. This is because, the product of an odd number of -1s is -1.

Concept 79:
The nth root of an expression is the same as raising the expression to the power of $\frac{1}{n}$.

Therefore, the nth root of a^m is: $\sqrt[n]{a^m} = (a^m)^{\frac{1}{n}} = a^{\frac{m}{n}}$.

1. Therefore, the square root of a term or expression is the same as raising the term or expression to the power of $\frac{1}{2}$.

2. Similarly, the cube root of a term or expression is the same as raising the term or expression to the power of $\frac{1}{3}$.

Concept 80:
The method of determining the root of a simple expression is as follows:

Elementary Algebra

1. Determine the root of the numerical coefficient.

2. Divide the exponent of each of every factor in the expression by the index of the root.

Concept 81:
The square root of common expressions:

1. $(a+b)^2 = a^2 + 2ab + b^2$
   Therefore the square root of $a^2 + 2ab + b^2 = \pm(a+b)$.

2. $(a-b)^2 = a^2 - 2ab + b^2$
   Therefore the square root of $a^2 - 2ab + b^2 = \pm(a-b)$.

3. $(a+b+c)^2 = a^2 + b^2 + c^2 + 2ab + 2bc + 2ac$
   Therefore the square root of $a^2 + b^2 + c^2 + 2ab + 2bc + 2ac = \pm(a+b+c)$

Similarly, we can make the following conclusions.

1. The cube root of $a^3 + b^3 + 3a^2b + 3ab^2 = (a+b)$.

2. The cube root of $a^3 - b^3 - 3a^2b + 3ab^2 = (a-b)$.

Question 12.1: Write down the square root of the following expressions:

1. $9x^4y^3$

2. $25a^6b^4$

3. $49c^2d^6$

4. $\dfrac{16x^{64}}{25}$

5. $\dfrac{4x^6}{16a^4}$

Question 12.2: Write down the cube roots of the following expressions:

1. $x^6 y^9$

2. $-a^6 b^3$

3. $8x^{27}$

4. $-27x^9$

5. $\dfrac{-b^{27}}{27}$

Question 12.3: Write down the square root of each of the following expressions using method of inspection:

1. $a^2 - 8a + 16$

2. $x^2 + 14x + 49$

3. $64 + 48x + 9x^2$

4. $25 - 30m + 9m^2$

5. $4a^2 b^4 - 12ab^2 c^5 + 9c^{10}$

13   Factorization

Concept 82:
When an algebraic term or an expression is a product of two or more factors, the process of determining these factors is called factorization or resolution into factors.

Concept 83:
For any algebraic term, the coefficient, each of the literals and its constituent powers are the factors.

This means that $6x^2y^3$ has $1, 2, 3, x, x^2, y, y^2$ and y^3 as factors.

Concept 84:
For a simple expression we use the following technique:

1. We find the common factors across the terms of the expression.

2. We divide each of the terms with these factors.

3. We enclose the quotients within brackets and multiply this expression. with the common factors.

For example, in the expression, $ax^2 + bx$, x is a common factor between ax^2 and bx.
Therefore we can write $ax^2 + bx = x(ax + b)$.
Therefore the factors of $ax^2 + bx$ are x and $(ax + b)$

We can also arrange the expression into factors if the terms can be grouped to identify the common factors. Let us consider an example to understand this well.

$x^2 + xy + xz + yz = x(x + y) + z(x + y)$
x is the common factor between the first two terms and z is the common factor between the last two terms.

On resolving the expression into factors, another common compound factor emerges between the $x(x + y)$ and $z(x + y)$, which is $(x + y)$.

Therefore, we can repeat the process of resolution one more time.

Let us put all the steps together to ensure that we get the flow of the thought.
$$x^2 + xy + xz + yz = x(x+y) + z(x+y) = (x+y)(x+z)$$

Concept 85:
The resolution of factors in binomial expressions is the next concept of interest. Let us start the discussion with a simple case of two compound expressions creating a trinomial expression.

Multiplying $(x+a)(x+b)$, we get $x^2 + x(a+b) + ab$
Therefore, $(x+a)(x+b) = x^2 + x(a+b) + ab$
Or the factors of $x^2 + x(a+b) + ab$ are $(x+a)(x+b)$.

For example, the process of resolving $x^2 + 8x + 15$ is as follows.
$$x^2 + 8x + 15 = (x+a)(x+b) = x^2 + x(a+b) + ab$$
Therefore $ab = 15$ and $(a+b) = 8$.
By inspection, we can see that $a = 5$ and $b = 3$
Therefore $x^2 + 8x + 15 = (x+5)(x+3)$

Concept 86:
The difference of two squares can be resolved into factors by using the following formula:
$$a^2 - b^2 = (a+b)(a-b)$$

The product of sum and difference of two terms is equal to the difference of their squares.

Conversely, the difference of two squares is equal to the product of the sum and difference of two quantities.

The sum and difference constitute the two factors of interest.

Concept 87:
Two more formulas of interest are:

1. $a^3 + b^3 = (a+b)(a^2 - ab + b^2)$

2. $a^3 - b^3 = (a-b)(a^2 + ab + b^2)$

The left hand sides of the equations represent the sum and difference of cubes. The right hand side tells us how to resolve them into factors.

Concept 88:
Using the results from the concepts we just saw, the ones dealing with factorization of sum and difference of squares and cubes respectively, we can factorize differences of larger powers. One such example is shown below.

$$(a^6 - b^6) = (a^3 + b^3)(a^3 - b^3) = (a+b)(a^2 - ab + b^2)(a-b)(a^2 + ab + b^2)$$

This is a good example to highlight how one can systematically approach problems in factorization.

Concept 89:
Let us now consider the generic case of a trinomial.
$$ax^2 + bx + c = (px + q)(rx + s) = prx^2 + x(ps + qr) + qs$$

Therefore, we have $pr = a$; $ps + qs = b$; and $qs = c$.

We can use these relationships to make intelligent conclusions about the factors and their coefficients p, q, r and s.

To start the process, check if a, b or c are prime numbers.

If a is a prime number, then either p or r or both are equal to 1. And the other variable is equal to a. We plug these numbers to determine the other coefficients and complete the solution.

Usually, a couple of intelligent guesses should do the trick.

Concept 90:
With a good grasp of techniques for factorization, we can look at determining the HCF of a set of algebraic expressions. HCF is defined as the expression of the highest dimension that divides a set of algebraic expressions.

Question 13.1: Resolve into factors

1. $x^2 + ax$

2. $2a^2 - 3a$

3. $a^3 - a^2$

4. $a^3 - a^2 b$

5. $3m^2 - 6mn$

6. $p^2 + 2p^2 q$

7. $p^2 12p^2 q$

8. $y^2 + xy$

9. $12x + 48x^2 y$

10. $10c^3 - 25c^4 d$

Question 13.2: Resolve into factors

1. $x^2 + xy + xz + yz$

2. $x^2 - xz + xy - yz$

3. $a^2 + 2a + ab + 2b$

4. $a^2 + ac + 4a + 4c$

5. $2a + 2x + ax + x^2$

6. $3q - 3p + pq - p^2$

7. $am - bm - am + bn$

8. $ab - by - ay + y^2$

9. $pq + qr - pr - r^2$

10. $2x^3 + 3 + 2x + 3x^2$

Question 13.3: Resolve into factors

1. $x^2 + 3x + 2$
2. $y^2 + 5y + 6$
3. $y^2 + 7y + 12$
4. $a^2 - 3a + 2$
5. $a^2 - a - 2$
6. $b^2 - 5b + 6$
7. $b^2 + 13b + 42$
8. $b^2 - 13b + 40$
9. $z^2 - 13z + 36$
10. $x^2 - 15x + 56$

Question 13.4: Resolve into factors

1. $2a^2 + 3a + 1$
2. $3a^2 + 4a + 1$
3. $4a^2 + 5a + 1$
4. $2a^2 + 5a + 2$
5. $3a^2 + 10a + 3$
6. $2a^2 + 7a + 3$
7. $5a^2 + 7a + 2$
8. $2a^2 + 9a + 10$
9. $2a^2 + 7a + 6$
10. $2x^2 + 9x + 4$

Elementary Algebra

Question 13.5:   Resolve into factors

1. $a^2 - 9$

2. $a^2 - 49$

3. $a^2 - 81$

4. $81 - 4x^2$

5. $a^6 b^8 c^4 - 9$

6. $8x^3 + 1$

7. $x^3 - 8z^3$

8. $27 + x^3$

9. $512a^3 - 1$

10. $x^6 - 27z^3$

14 Quadratic Equations

Concept 91:
An equation of the second degree is also known as the quadratic equation. The general form of a quadratic equation is $ax^2 + bx + c = 0$.

Here is a mathematical trivia. A quadratic equation of the form $ax^2 + bx + c = 0$ is also known as adfected quadratic; while a quadratic equation of the form $ax^2 + c = 0$ is also known as pure quadratic. We will drop the reference to the term "adfected" and simple refer to these equations as general quadratics.

We will commence our discussion of quadratic equations with pure quadratics.

Concept 92:
A pure quadratic equation is of the form $ax^2 + c = 0$. We are solving for the value of x given x^2. In other words, we are dealing with finding the roots of an equation. Therefore, it is important for us to quickly recall the rule of signs for determining roots.

An even root of any positive term or quantity can be either positive or negative. This is because product of two positive terms or two negative terms is always positive. Therefore, the root of a quadratic—an even root, can be positive or negative.

If $x^2 = 25$; then $x = +5$ or $x = -5$ will satisfy the equation. This is represented by $x = \pm 5$ and is read as "x equals plus or minus five".

The roots of a pure quadratic of the type $ax^2 + c = 0$ are $x = \pm\sqrt{\dfrac{-c}{a}}$

This formula is usually not memorized. We simply transpose the numbers and constant terms to the right hand side of the equation and collect all the terms with second power of the unknown to the left hand side of the equation.

The solution to the determining the root of the unknown then becomes obvious and straightforward.

For example:

$$\frac{9}{x^2 - 27} = \frac{25}{x^2 - 11}$$

$$9(x^2 - 11) = 25(x^2 - 27)$$

$$9x^2 - 99 = 25x^2 - 675$$

$$16x^2 = 576$$

$$x^2 = 36$$

Therefore, $x = \pm 6$

The process of arriving at the solution to a pure quadratic from first principles is simple and straightforward.

Concept 93:

Although a few expressions are not strictly pure quadratics, we can still use the above technique in handling general quadratic equations.

Consider $(x + a)^2 = b$

Then we can conclude that $x + a = \pm\sqrt{b}$

Therefore the solution to the equation is $x = -a \pm \sqrt{b}$

In other words, if we can convert a general quadratic into an equation of the form $(x + a)^2$ by completing the square, we can solve the roots like we did before.

Concept 94:

Let us look at a general quadratic equation:

$$ax^2 + bx + c = 0$$

Dividing both sides of the equation by the coefficient of x^2:

$$x^2 + \frac{b}{a}x + \frac{c}{a} = 0$$

Transposing $\dfrac{c}{a}$ to the right hand side of the equation:

$$x^2 + \frac{b}{a}x = -\frac{c}{a}$$

Let us complete the square on the left hand side of the equation.

Add $\left(\dfrac{b}{2a}\right)^2$ to both sides:

$$x^2 + 2\frac{b}{2a}x + \left(\frac{b}{2a}\right)^2 = -\frac{c}{a} + \left(\frac{b}{2a}\right)^2 = \frac{b^2}{4a^2} - \frac{4ac}{4a^2}$$

$$\left(x + \frac{b}{2a}\right)^2 = \frac{b^2 - 4ac}{4a^2}$$

$$x + \frac{b}{2a} = \pm\sqrt{\frac{b^2 - 4ac}{(2a)^2}} = \pm\frac{\sqrt{b^2 - 4ac}}{2a}$$

$$x = \frac{-b \pm \sqrt{b^2 - 4ac}}{2a}$$

These are the two roots of a general quadratic equation.

Question 14.1: Solve the following equations:

1. $7(x^2 - 7) = 6x^2$

2. $(x + 8)(x - 8) = 0$

3. $(7 + x)(7 - x) = 0$

4. $\dfrac{x^2 + 8}{x^2 + 20} = \dfrac{1}{2}$

5. $\dfrac{11}{3 - x} = 4(x + 3)$

6. $\dfrac{x(3x + 5) + 21}{(3x - 2)(2x + 3)} = 1$

7. $x^2 + 2x = 8$

8. $x^2 + 6x = 40$

9. $x^2 + 35 = 12x$

10. $x^2 + 15x - 34 = 0$

Question 14.2:   Solve the following equations:

1. $3x^2 + 2x = 21$

2. $5x^2 = 8x + 21$

3. $6x^2 - x - 1 = 0$

4. $3 - 11x = 4x^2$

5. $21x^2 = 2x + 3$

6. $10 + 23x + 12x^2 = 0$

7. $15x^2 - 6x = 9$

8. $4x^2 - 17x = 15$

9. $8x^2 - 19x - 15 = 0$

10. $7(x + 2a)^2 + 3a^2 = 5a(7x + 23a)$

Question 14.3:   Solve the following equations:

1. $x^2 + 2x - 3 = 0$

2. $x^2 - 2x - 1 = 0$

3. $x^2 - 3x - 5 = 0$

4. $3x^2 - 2x - 1 = 0$

5. $2x^2 - 9x - 4 = 0$

6. $3x^2 + 7x - 6 = 0$

7. $4x^2 - 3x - 14 = 0$

8. $6x^2 - 7x - 3 = 0$

9. $12x^2 - 23x + 10 = 0$

10. $x^2 - 9x - 90 = 0$

15   Ratio and Proportion

Concept 95:
Ratio is simply a relationship between two similar quantities. The nature of comparison is about what part of one is the other; or what multiple of one is the other. This is written as $a : b$. The first part a is called antecedent and the other part b is called consequent.

The ratio is said to be of greater inequality, equality or lesser inequality when antecedent is greater than, equal to or less than the consequent respectively.

Concept 96:
A ratio $a : b$ is the same as the fraction $\dfrac{a}{b}$.

Concept 97:
The laws of fraction are applicable to ratios as well.

$$\frac{a}{b} = \frac{ma}{mb} \text{ for all } m \neq 0$$

Therefore $a : b = ma : mb$ for all $m \neq 0$

In other words, the value of a ratio remains unaltered if the antecedent and consequent of a ratio are divided by the same quantity.

Concept 98:
Let us consider two ratios $a : b$ and $x : y$

$$a : b = \frac{a}{b} = \frac{ay}{by} \text{ (value is not altered if we multiply and divide}$$
by the same quantity)

$$x : y = \frac{x}{y} = \frac{bx}{by} \text{ (value is not altered if we multiply and divide}$$
by the same quantity)

Therefore

1. $a : b > x : y$ when $ay > bx$

2. $a : b = x : y$ when $ay = bx$

3. $a : b < x : y$ when $ay < bx$

The above relationship is also known as rule of cross products.

Concept 99:
If $a : b$, $p : q$ and $x : y$ are given ratios, then $apx : bqy$ is called the compounded ratio. We simply take the product of antecedents and consequents to determine the compounded ratio of two or more ratios.

Concept 100:
When $x : y$ is compounded by itself, we get $x^2 : y^2$. This is called the duplicate ratio of $x : y$.
And, $x^3 : y^3$ is called triplicate of the ratio $x : y$.
$\sqrt{x} : \sqrt{y}$ is called sub-duplicate ratio of $x : y$

Concept 101:
If $a : b = c : d = e : f$, then $ax + by + cz : bx + dy + ez = a : b = c : d = e : f$

In other words, when a series of fractions or ratios are equal to one another, then each of the fractions is equal to the sum of all numerators divided by sum of all denominators.

Concept 102:
When two ratios are equal, then the four quantities are said to be in proportion.

If $a : b = c : d$, then a, b, c and d are in proportion.
This is written as $a : b :: c : d$, and read out as a is to b is as is to c is to d.

a and d are called extremes and b and c are known as means.

It may be noted that when four quantities are in proportion, then product of means is equal to product of extremes.

Concept 103:
Three quantities a, b and c are said to be in continued proportion if
$a : b :: b : c$ or $\dfrac{a}{b} = \dfrac{b}{c}$
The product of means = Product of extremes
$b^2 = ac$

b is called the mean proportion of a and c.

Concept 104:
If three quantities are in continued proportion, then the ratio of the first and third quantity is the duplicate ratio of first and second quantity.

If $a : b : c$, then $a : c :: a^2 : b^2$

Concept 105:
If $a : b :: c : d$ and $e : f :: g : h$, then $ae : bf :: cg : dh$
As a consequence, if $a : b :: c : d$ and $b : x :: d : y$, then $a : x :: c : y$

Concept 106:

1. If $a : b :: c : d$ then $b : a :: d : c$

2. If $a : b :: c : d$ then $a : c :: b : d$

3. If $a : b :: c : d$ then $(a + b) : b :: (c + d) : d$

4. If $a : b :: c : d$ then $(a - b) : b :: (c - d) : d$

5. If $a : b :: c : d$ then $(a + b) : (a - b) :: (c + d) : (c - d)$

Concept 107:
If $a : b :: c : d$, then

1. $ma : mb :: nc : nd$

2. $ma : nb :: mc : nd$

3. $an : bn :: cn : dn$

4. $pan : qbn :: pcn : qdn$

Question 15.1:   If $3(4x - 5y) = 2x - y$, find the ratio of $x : y$.

Question 15.2:   If $\dfrac{5a + 3b}{4a + 5b} = \dfrac{2}{3}$, find the ratio of $a : b$.

Question 15.3:   Two numbers are in the ratio of $5 : 7$. If 9 is added to each of them, the resulting numbers are in the ratio of $4 : 5$. What are the number?

Question 15.4:   If $\dfrac{a}{b} = \dfrac{7}{6}$, find the value of $(3a + 5b) : (7b - 5a)$.

Question 15.5:   If $\dfrac{2x}{3y} = \dfrac{5}{4}$, find the value of $(8x - 7y) : (y + 8x)$

Question 15.6:   If $16a = 25b$, find the duplicate ratio of $a : b$.

Question 15.7:   If $25x = 9y$, find the subduplicate ratio of $x : y$.

Question 15.8:   If $\dfrac{x}{cm - bn} = \dfrac{y}{cl - an} = \dfrac{z}{bl - am}$, show that $ax - by + cz = 0$

Question 15.9:   Solve the following equations:

1. $(3x - 5) : (5x - 11) = 2 : 3$

2. $(2x + 1) : (x + 5) = (6x - 7) : (3x + 5)$

3. $(3x - 2) : (x + 2) = (5x - 2) : (x + 8)$

4. $x : y = 3 : 4 = (x + y) : (3x + 1)$

Question 15.10:   If a, b, c be proportionals. show that:

1. $a - b : b - c = b : c$

2. $a + b : a - b = b + c : b - c$

3. $a + b : b + c = a : b$

4. $ma + nb : mb + nc = ma - nb : mb - nc$

5. $a^2 + b^2 : (a + b)^2 = b^2 + c^2 : (b + c)^2$

16 Closing Thoughts

In this book, we have covered over 100 concepts and over 500 problems and solutions. As we come to the end of the book, we would like to summarize a few critical inputs:

1. Mathematics is a universe filled with concepts.

2. To solve problems, make use of the concepts – both fundamental and derived ones.

3. Therefore, it is important for us to elaborate the nature of our thought process while solving problems. We have to be systematic.

4. There is no victory if we skip steps and jump ahead a few steps without outlining our assumptions and rationale.

5. All deductions must be made from fundamental principles or equations that are well established.

6. The sign is an important part of the solution. The roots to a quadratic, for example, will always be plus or minus. We need to pay close attention to details.

7. Above all, Mathematics is not a chore. It is a fun filled exercise. It is important to have fun and enjoy every step of the problem solving process.

It has been our sincere effort to research several sources and texts to bring together this work on Elementary Algebra. The purpose is to enable a strong conceptual foundation in the language of Algebra. If this book succeeds to set a spark of excitement in the reader and compels him or her to explore this universe a bit further, this book would have achieved its objective.

Cheers,
Chandramouli Mahadevan
On behalf of Team Astrarka

17 Solutions

Solutions to Chapter 2

Question 2.1: if $a = 5, b = 4, c = 1, x = 3, y = 12, z = 2$, find the values of

1. $2a = 2 \times a = 2 \times 5 = 10$
2. $a^2 = a \times a = 5 \times 5 = 25$
3. $3z = 3 \times z = 3 \times 2 = 6$
4. $z^3 = z \times z \times z = 2 \times 2 \times 2 = 8$
5. $c^4 = c \times c \times c \times c = 1 \times 1 \times 1 \times 1 = 1$
6. $4c = 4 \times c = 4 \times 1 = 4$
7. $4b^2 = 4 \times b \times b = 4 \times 4 \times 4 = 64$
8. $c^3 = c \times c \times c = 1 \times 1 \times 1 = 1$
9. $x^3 = x \times x \times x = 3 \times 3 \times 3 = 27$
10. $3x = 3 \times x = 3 \times 3 = 9$
11. $7y^2 = 7 \times 12^2 = 7 \times 144 = 1008$
12. $8a^3 = 8 \times 5^3 = 8 \times 125 = 1000$
13. $6z^5 = 6 \times 2^5 = 6 \times 32 = 192$
14. $5z^6 = 5 \times 2^6 = 5 \times 64 = 320$
15. $7c^6 = 7 \times 1^6 = 7 \times 1 = 7$

Question 2.2: if $a = 6, p = 4, q = 7, r = 5, x = 1$, find the value of

1. $ap = 6 \times 4 = 24$
2. $3pq = 3 \times 4 \times 7 = 84$

3. $3qx = 3 \times 7 \times 1 = 21$

4. $5p^3 = 5 \times 4^3 = 320$

5. $8aqx = 8 \times 6 \times 7 \times 1 = 336$

6. $pqr = 4 \times 7 \times 5 = 140$

7. $8aqr = 8 \times 6 \times 7 \times 5 = 1680$

8. $7qrx = 7 \times 7 \times 5 \times 1 = 245$

9. $2apx = 2 \times 6 \times 4 \times 1 = 48$

10. $7x^4 = 7 \times 1^4 = 7$

11. $3p^4 = 3 \times 4^4 = 768$

12. $8r^4 = 8 \times 5^4 = 5000$

13. $9apqx = 9 \times 6 \times 4 \times 7 \times 1 = 1512$

14. $6x^7 = 6 \times 1^7 = 6$

15. $x^{10} = 1^{10} = 1$

Question 2.3: if $a = 4, b = 1, c = 3, f = 5, g = 7, h = 0$, find the value of

1. $3f + 5h - 7b = 3 \times 5 + 5 \times 0 - 7 \times 1 = 8$

2. $7c - 9h + 2a = 7 \times 3 - 9 \times 0 + 2 \times 4 = 29$

3. $4g - 5c - 9b = 4 \times 7 - 5 \times 3 - 9 \times 1 = 4$

4. $3g - 4h + 7c = 3 \times 7 - 4 \times 0 + 7 \times 3 = 42$

5. $3f - 2g - b = 3 \times 5 - 2 \times 7 - 1 = 0$

6. $9b - 3c + 4h = 9 \times 1 - 3 \times 3 + 4 \times 0 = 0$

7. $3a - 9b + c = 3 \times 4 - 9 \times 1 + 3 = 6$

8. $2f - 3g + 5a = 2 \times 5 - 3 \times 7 + 5 \times 4 = 9$

9. $3c - 4a + 7b = 3 \times 3 - 4 \times 4 + 7 \times 1 = 0$

10. $3f + 5h - 2c - 4b + a = 3 \times 5 + 5 \times 0 - 2 \times 3 - 4 \times 1 + 4 = 9$

11. $6h - 7b - 5a - 7f + 9g = 6 \times 0 - 7 \times 1 - 5 \times 4 - 7 \times 5 + 9 \times 7 = 1$

12. $7c + 5b - 4a + 8h + 3g = 7 \times 3 + 5 \times 1 - 4 \times 4 + 8 \times 0 + 3 \times 7 = 31$

13. $9b + a - 3g + 4f + 7h = 9 \times 1 + 4 - 3 \times 7 + 4 \times 5 + 7 \times 0 = 12$

14. $fg + gh - ab = 5 \times 7 + 7 \times 0 - 4 \times 1 = 31$

15. $gb - 3hc + fb = 7 \times 1 - 3 \times 0 \times 3 + 5 \times 1 = 12$

16. $fh + hb - 3hc = 5 \times 0 + 0 \times 1 - 3 \times 0 \times 3 = 0$

17. $f^2 - 3a^2 + 2c^3 = 5^2 - 3 \times 4^2 + 2 \times 3^3 = 31$

18. $b^3 - 2h^3 + 3a^2 = 1^3 - 2 \times 0^3 + 3 \times 4^2 = 51$

19. $3b^2 - 2b^3 + 4h^2 - 2h^4 = 3 \times 1^2 - 2 \times 1^3 + 4 \times 0^2 - 2 \times 0^4 = 1$

Solutions to Chapter 3

Question 3.1: Find the sum of

1. $2a + 3a + 6a + a + 4a = 16a$

2. $4x + x + 5x + 6x + 8x = 22x$

3. $6b + 11b + 8b + 9b + 5b = 39b$

4. $6c + 7c + 3c + 16c + 18c + 101c = 151c$

5. $2p + p + 4p + 7p + 6p + 12p = 32p$

6. $d + 9d + 3d + 7d + 4d + 6d + 10d = 40d$

7. $-2x - 6x - 10x - 8x = -26x$

8. $-3b - 13b - 19b - 5b = -40b$

9. $-y - 4y - 2y - 6y - 4y = -17y$

10. $-17c - 34c - 9c - 6c = -66c$

11. $-21y - 5y - 3y - 18y = -47y$

12. $-4m - 13m - 17m - 59m = -93m$

13. $3x - 10x - 7x + 12x + 2x = 0$

14. $8ab - 6ab + 5ab - 4ab = 3ab$

15. $2xy - 4xy - 3xy + xy + 7xy = 3xy$

Question 3.2: Find the value of

1. $-9a^2 + 11a^2 + 3a^2 - 4a^2 = a^2$

2. $b^3 - 2b^3 + 7b^3 - 9b^3 = -3b^3$

3. $-11a^3 + 3a^3 - 8a^3 - 7a^3 + 2a^3 = -21a^3$

4. $a^2b^2 - 7a^2b^2 + 8a^2b^2 + 9a^2b^2 = 11a^2b^2$

5. $a^2x - 11a^2x + 3a^2x - 2a^2x = -9a^2x$

6. $2p^3q^3 - 31p^3q^3 + 17p^3q^3 = -12p^3q^3$

7. $7m^4n - 15m^4n + 3m^4n = -5m^4n$

8. $9abcd - 11abcd - 41abcd = 39abcd$

9. $13pqx - 5pqx - 19pqx = -11pqx$

10. $2x^3 - 3x^3 - 6x^3 - 9x^3 = -16x^3$

Question 3.3: Find the sum of:

1. $(3a + 2b - 5c); (-4a + b - 7c); (4a - 3b + 6c)$
   $(3a + 2b - 5c) + (-4a + b - 7c) + (4a - 3b + 6c)$
   Grouping like terms together, we get the desired answer.
   $(3a + 2b - 5c) + (-4a + b - 7c) + (4a - 3b + 6c)$
   $= 3a + 0b - 6c = 3a - 6c$

2. $(3x + 2y + 6z); (x - 3y - 3z); (2x + y - 3z)$
   $(3x + 2y + 6z) + (x - 3y - 3z) + (2x + y - 3z)$
   Grouping like terms together, we get the desired answer.
   $= (3x + x + 2x) + (2y - 3y + y) + (6z - 3z - z)$
   $= 6x + 0y + 2z = 6x + 2z$

3. $4p + 3q + 5r; -2p + 3q - 8r; p - q + r$
$(4p + 3q + 5r) + (-2p + 3q - 8r) + (p - q + r)$
Grouping like terms together, we get the desired answer.
$(4p - 2p + p) + (3q + 3q - q) + (5r - 8r + r)$
$= 3p + 5q - 2r$

4. $7a - 5b + 3c; 11a + 2b - c; 16a + 5b - 2c$
$(7a - 5b + 3c) + (11a + 2b - c) + (16a + 5b - 2c)$
Grouping like terms together, we get the desired answer.
$(7a + 11a + 16a) + (-5b + 2b + 5b) + (3c - c - 2c)$
$= 34a + 2b$

Now that we have got the drift of the process, let us write down the sum through inspection of like terms.

5. $8l - 2m + 5n; -6l + 7m + 4n; -l - 4m - 8n$
$8l - 2m + 5n - 6l + 7m + 4n - l - 4m - 8n$
$= l + m + n$

6. $5a - 7b + 3c - 4d; 6b - 5c + 3d; b + 2c - d$
$5a - 7b + 3c - 4d + 6b - 5c + 3d + b + 2c - d$
$= 5a - 2d$

7. $2a + 4b - 5x; 2b - 5x; -3a + 2y; -6b + 8x + y$
$2a + 4b - 5x + 2b - 5x - 3a + 2y - 6b + 8x + y$
$= -a - 2x + 3y$

8. $7x - 5y - 7z; 4x + y; 5z; 5x - 3y + 2z$
$7x - 5y - 7z + 4x + y + 5z + 5x - 3y + 2z$
$= 16x - 7y - 5z$

9. $5 - x - y; 7 + 2x; 3y - 2z; -4 + x - 2y$
$5 - x - y + 7 + 2x + 3y - 2z - 4 + x - 2y$
$= 8 + 2x - 2z$

10. $25a - 15b + c; 4c - 10b + 13a; a - c + 20b$
$25a - 15b + c + 4c - 10b + 13a + a - c + 20b$
$= 39a - 5b + 4c$

11. $2a - 3b - 2c + 2x; 5x + 3b - 7c; 9c - 6x - 2a$
$2a - 3b - 2c + 2x + 5x + 3b - 7c + 9c - 6x - 2a$
$= x$

12. $3a - 5c + 2b - 2d; b + 2d - a; 5c + 3f + 3e - 2a - 3b$
   $3a - 5c + 2b - 2d + b + 2d - a + 5c + 3f + 3e - 2a - 3b$
   $= 3e + 3f$

13. $p - q + 7r; 6q + r - p; q - 3p - r; 6q - 7p$
   $p - q + 7r + 6q + r - p + q - 3p - r + 6q - 7p$
   $= -10p + 6q + 7r$

14. $17ab - 13kl - 5xy; 7xy; 12kl - 5ab; 3xy - 4kl - ab$
   $17ab - 13kl - 5xy + 7xy + 12kl - 5ab + 3xy - 4kl - ab$
   $= 11ab - 5kl + 3xy$

15. $2ax - 3by - 2cz; 2by - ax + 7cz; ax - 4cz + 7by; cz - 6by$
   $2ax - 3by - 2cz + 2by - ax + 7cz + ax - 4cz + 7by + cz - 6by$
   $= 2ax + 2cz$

Question 3.4: Find the sum of the following expressions:

1. $x^2 + 3xy - 3y^2; -3x^2 + xy + 2y^2; 2x^2 - 3xy + y^2$
   $x^2 + 3xy - 3y^2 - 3x^2 + xy + 2y^2 + 2x^2 - 3xy + y^2$
   $= 0$

2. $2x^2 - 2x + 3; -2x^2 + 5x + 4; x^2 - 2x - 6$
   $2x^2 - 2x + 3 - 2x^2 + 5x + 4 + x^2 - 2x - 6$
   $= x^2 + x + 1$

3. $5x^3 - x^2 + x - 1; 2x^2 + 5x + 4; x^2 - 2x - 6$
   $5x^3 - x^2 + x - 1 + 2x^2 + 5x + 4 + x^2 - 2x - 6$
   $= 5x^3 + 2x^2 + 4x - 3$

4. $a^3 - a^2b + 5ab^2 + b^3; -a^3 - 10ab^2 + b^3; 2a^b + 5ab^2 - b^3$
   $a^3 - a^2b + 5ab^2 + b^3 - a^3 - 10ab^2 + b^3 + 2a^b + 5ab^2 - b^3$
   $= b^3 - 11a^b + 10ab^2 + 2a^b$

5. $3x^3 - 9x^2 - 11x + 7; 2x^3 - 5x^2 + 2; 5x^3 + 15x^2 - 7x; 8x - 9$
   $3x^3 - 9x^2 - 11x + 7 + 2x^3 - 5x^2 + 2 + 5x^3 + 15x^2 - 7x + 8x - 9$
   $= 10x^3 + x^2 - 10x$

6. $4m^3 + 2m^2 - 5m + 7; 3m^3 + 6m^2 - 2; -5m^2 + 3m; 2m - 6$
   $4m^3 + 2m^2 - 5m + 7 + 3m^3 + 6m^2 - 2 - 5m^2 + 3m + 2m - 6$
   $= 7m^3 + 3m^2 - 1$

7. $ax^3 - 4bx^2 + cx; 3bx^2 - 2cx - d; bx^2 + 2d; 2ax^3 + d$
   $ax^3 - 4bx^2 + cx + 3bx^2 - 2cx - d + bx^2 + 2d + 2ax^3 + d$
   $= 3ax^3 - cx + 2d$

8. $py^2 - 9qy + 7r; -2py^2 + 3qy - 6r; 7qy - 4r; 3py^2$
   $py^2 - 9qy + 7r - 2py^2 + 3qy - 6r + 7qy - 4r + 3py^2$
   $= 2py^2 + qy - 3r$

9. $5y^3 + 20y^2 + 3y - 1; -2y + 5 - 7y^2; -3y^2 - 4 + 2y^3 - y$
   $5y^3 + 20y^2 + 3y - 1 - 2y + 5 - 7y^2 - 3y^2 - 4 + 2y^3 - y$
   $= 7y^3 + 10y^2$

10. $2 - a + 8a^2 - a^3; 2a^3 - 3a^2 + 2a - 2; -3a + 7a^3 - 5a^2$
    $2 - a + 8a^2 - a^3 + 2a^3 - 3a^2 + 2a - 2 - 3a + 7a^3 - 5a^2$
    $= 8a^3 - 2a$

11. $1 + 2y - 3y^2 - 5y^3; -1 + 2y^2 - y; 5y^3 + 3y^2 + 4$
    $1 + 2y - 3y^2 - 5y^3 - 1 + 2y^2 - y + 5y^3 + 3y^2 + 4$
    $= 2y^2 + y + 4$

12. $c^7 - 2c^5 + 11c^6; -2c^7 - 2c^6 + 5c^5; 4c^6 - 10c^5; 4c^7 - c^6$
    $c^7 - 2c^5 + 11c^6 - 2c^7 - 2c^6 + 5c^5 + 4c^6 - 10c^5 + 4c^7 - c^6$
    $= 3c^7 + 12c^6 - 7c^5$

13. $4h^3 - 7 + 3h^4 - 2h; 7h - 3h^3 + 2 - h^4; 2h^4 + 2h^3 - 5$
    $4h^3 - 7 + 3h^4 - 2h + 7h - 3h^3 + 2 - h^4 + 2h^4 + 2h^3 - 5$
    $= 5h^4 + 3h^3 + 5h - 10$

14. $x^2 + 2xy + 3y^2; 3z2 + 2yz + y^2; x^2 + 3z^2 + 2xz; z2 - 3xy - 3yz; xy + xz + yz - 6z^2 - 4y^2 - 2x^2$
    $x^2 + 2xy + 3y^2 + 3z2 + 2yz + y^2 + x^2 + 3z^2 + 2xz + z2 - 3xy - 3yz + xy + xz + yz - 6z^2 - 4y^2 - 2x^2$
    $= 3xz$

Question 3.5: Subtract

If we find a - sign in front of a bracket, the sign of every term inside the bracket is changed; we then add all like terms together. This is the general method of solving subtraction problems.

1. $a + 2b - c$ from $2a + 3b + c$
   $(2a + 3b + c) - (a + 2b - c) = a + b + 2c$

2. $2a - b + c$ from $3a - 5b - c$
$(3a - 5b - c) - (2a - b + c) = a - 4b - 2c$

3. $3x + y - z$ from $x - 4y + 3z$
$(x - 4y + 3z) - (3x + y - z) = -2x - 5y + 4z$

4. $x + 8y + 8z$ from $10x - 7y - 6z$
$(10x - 7y - 6z) - (x + 8y + 8z) = 9x - 15y - 14z$

5. $-m - 3n + p$ from $-2m + n - 3p$
$(-2m + n - 3p) - (-m - 3n + p) = -m + 4n - 4p$

6. $3p - 2q + r$ from $4p - 7q + 3r$
$(4p - 7q + 3r) - (3p - 2q + r) = p - 5q + 2r$

7. $a - 7b - 3c$ from $-4a + 3b + 8c$
$(-4a + 3b + 8c) - (a - 7b - 3c) = -5a + 10b + 11c$

8. $-a - b - 9c$ from $-a + b - 9c$
$(-a + b - 9c) - (-a - b - 9c) = 2b$

9. $3x - 5y - 7z$ from $2x + 3y - 4z$
$(2x + 3y - 4z) - (3x - 5y - 7z) = -x + 8y + 11z$

10. $-4x - 2y + 11z$ from $-x + 2y - 13z$
$(-x + 2y - 13z) - (-4x - 2y + 11z) = 3x + 4y - 24z$

11. $-2x - 5y$ from $x + 3y - 2z$
$(x + 3y - 2z) - (-2x - 5y) = 3x + 8y - 2z$

12. $3x - y - 8z$ from $x + 2y$
$(x + 2y) - (3x - y - 8z) = -2x + 3y + 8z$

13. $m - 2n - p$ from $m + 2n$
$(m + 2n) - (m - 2n - p) = 4n + p$

14. $2p - 3q - r$ from $2q - 4r$
$(2q - 4r) - (2p - 3q - r) = -2p + 5q - 3r$

15. $ab - 2cd - ac$ from $-ab - 3cd + 2ac$
$(-ab - 3cd + 2ac) - (ab - 2cd - ac) = -2ab - cd + 3ac$

16. $3ab + 6cd - 3ac - 5bd$ from $3ab + 5cd - 4ac - 6bd$
$(3ab + 5cd - 4ac - 6bd) - (3ab + 6cd - 3ac - 5bd) = -cd - ac - bd$

17. $-xy + yz - zx$ from $2xy + zx$
$(2xy + zx) - (-xy + yz - zx) = 3xy + 2zx - yz$

18. $-2pq - 3qr + 4rs$ from $qr - 4rs$
$(qr - 4rs) - (-2pq - 3qr + 4rs) = 2p + 4qr - 8rs$

19. $-mn + 11np$ from $-11np$
$(-11np) - (-mn + 11np) = mn - 22np$

20. $-x^3 + 3x^2 - x$ from $x^3 - 3x^2 + x$
$(x^3 - 3x^2 + x) - (-x^3 + 3x^2 - x) = 2x^3 - 6x^2 + 2x$

Question 3.6: When $x = 2$, $y = 3$, $z = 4$, find the value of sum of $5x^2$, $-3xy$ and z^2. Also find the value of $3z^x + 3x^y$.

Given: $x = 2, y = 3, z = 4$
1. Sum of $5x^2$, $-3xy$ and z^2
$5x^2 + -3xy + z^2 = 5 \times 2^2 - 3 \times 2 \times 3 + 4^2$
$= 20 - 18 + 16 = 18$

2. $3z^x + 3x^y$
$= 3 \times 4^2 + 3 \times 2^3$
$= 48 + 24 = 72$

Question 3.7: Add together $3ab + bc - ca, -ab + ca, ab - 2bc + 5ca$. From the sum, take away $5ca + bc - ab$.

$3ab + bc - ca + -ab + ca + ab - 2bc + 5ca - (5ca + bc - ab)$
$= 4ab - 2bc$

Question 3.8: Subtract the sum of $x - y + 3z$ and $-2y - 2z$ from the sum of $2x - 5y - 3z$ and $-3x + y + 4z$.

1. $x - y + 3z + -2y - 2z = x - 3y + z$
2. $2x - 5y - 3z + -3x + y + 4z = -x - 4y + z$
3. $-x - 4y + z - (x - 3y + z) = -2x - y$

Question 3.9: Simplify a). $3b - 2b^2 - (2b - 3b^2)$. b). $3a - 2b - (2b + a) - (a - 5b)$

a. $3b - 2b^2 - (2b - 3b^2) = b + b^2$
b. $3a - 2b - (2b + a) - (a - 5b) = a + b$

Question 3.10: Subtract $8c^2 + 8c - 2$ from $c^3 - 1$.

$(c^3 - 1) - (8c^2 + 8c - 2) = c^3 - 8c^2 - 8c + 1$

Question 3.11: Add together $3a^2 - 7a + 5$ and $2a^3 + 5a - 3$, and diminish the result by $3a^2 + 2$.

1. $3a^2 - 7a + 5 + 2a^3 + 5a - 3 = 2a^3 + 3a^2 - 2a + 2$
2. $2a^3 + 3a^2 - 2a + 2 - (3a^2 + 2) = 2a^3 - 2a$

Question 3.12: Subtract $2b^2 - 2$ from $-2b + 6$, and increase the result by $3b - 7$.

1. $-2b + 6 - (2b^2 - 2) = -2b^2 - 2b + 8$
2. $-2b^2 - 2b + 8 + 3b - 7 = -2b^2 + b + 1$

Question 3.13: Find the sum of $3x^2 - 4x + 8$, $2x - 3 - x^2$ and $2x^2 - 2$, and subtract the result from $6x^2 + 3$.

1. $3x^2 - 4x + 8 + 2x - 3 - x^2 + 2x^2 - 2 = 5x^2 - 2x - 3$
2. $(6x^2 + 3) - (5x^2 - 2x - 3) = x^2 + 2x + 6$

Question 3.14: What expression must be added to $5a^2 - 3a + 12$ to produce $9a^2 - 7$?

$5a^2 - 3a + 12 + \text{REQUIRED-EXPRESSION} = 9a^2 - 7$
$\text{REQUIRED-EXPRESSION} = 9a^2 - 7 - (5a^2 - 3a + 12) = 4a^2 + 3a - 19$

Question 3.15: Find the sum of $2x$, $-x^3$, $3x^2$, 2, $-5x$, -4, $3x^3$, $-5x^2$, 8; and arrange the result in ascending powers of x.

1. $2x + -x^3 + 3x^2 + 2 + -5x + -4 + 3x^3 + -5x^2 + 8$
$= 2x^3 - 2x^2 - 3x + 6$
2. Required answer : $6 - 3x - 2x^2 + 2x^3$

Question 3.16: From what expression must the sum of $5a^2 - 2$, $3a + a^2$ and $7 - 2a$ be subtracted to produce $3a2 + a - 5$.

1. $5a^2 - 2 + 3a + a^2 + 7 - 2a = 6a^2 + a + 5$
2. $(6a^2 + a + 5) - (\text{REQUIRED-EXPRESSION}) = 3a2 + a - 5$
Therefore, REQUIRED-EXPRESSION
$= (6a^2 + a + 5) - (3a2 + a - 5) = 3a^2 + 10$

Question 3.17: When $x = 6$, find the numberical value of the sum of $1 - x + x^2, 2x^2 - 1$

$1 - x + x^2 + 2x^2 - 1 = 1 - 6 + 36 + 72 - 1 = 102$

Question 3.18: Find the value of $6ax + (2by - cz) - (2ax - 3by + 4cz) - (cz + ax)$, when $a = 0, b = 1, c = 2, x = 8, y = 3, z = 4$.

$6ax + (2by - cz) - (2ax - 3by + 4cz) - (cz + ax)$
$= +(2by - cz) - (-3by + 4cz) - (cz)$ when $a = 0$
$= (6 - 8) - (-9 + 32) - (8) = -2 - 23 - 8 = -33$

Question 3.19: Subtract the sum of $x^3 - 3x^2, 2x^2 - 7x, 8x - 2, 5 - 3x^3, 2x^3 - 7$ from $x^3 + x^2 + x + 1$.

1. $x^3 - 3x^2 + 2x^2 - 7x + 8x - 2 + 5 - 3x^3 + 2x^3 - 7 = -x^2 + x - 4$
2. $x^3 + x^2 + x + 1 - (-x^2 + x - 4) = x^3 + 2x^2 + 5$

Question 3.20: What expression must be taken from the sum of $p^4 - 3p^3, 2p + 8, 2p^2, 2p^3 - 3p^4$, in order to produce $4p^4 - 3$.

1. $p^4 - 3p^3 + 2p + 8 + 2p^2 + 2p^3 - 3p^4 = -2p^4 - p^3 + 2p^2 + 2p + 8$
2. $-2p^4 - p^3 + 2p^2 + 2p + 8 - \text{REQUIRED-EXPRESSION} = 4p^4 - 3$
Therefore, REQUIRED-EXPRESSION
$= -2p^4 - p^3 + 2p^2 + 2p + 8 - (4p^4 - 3)$
$= -6p^4 - p^3 + 2p^2 + 2p + 11$

Question 3.21: What is the result when $-3x^3 + 2x^2 - 11x + 5$ is subtracted from zero.

$0 - (-3x^3 + 2x^2 - 11x + 5) = 3x^3 - 2x^2 + 11x - 5$

Question 3.22: By how much does $b + c$ exceed $b - c$?
It exceeds by $(b + c) - (b - c)$ or $2c$.

Question 3.23: Find the algebraic sum of three times the square of x, twice the cube of x, $-x^3 + x - 2x^2$, and $x^3 - x - x^2 + 1$.

1. $3 \times x^2 = 3x^2$
2. $2 \times x^3 = 2x^3$
3. Required Sum : $3x^2 + 2x^3 + (-x^3 + x - 2x^2) + (x^3 - x - x^2 + 1)$
$= 2x^3 + 1$

Question 3.24: Take $p^2 - q^2$ from $3pq - 4q^2$, and add the remainder to the sum of $4pq - p^2 - 3q^2$ and $2p^2 + 6q^2$.

1. $(3pq - 4q^2) - (p^2 - q^2) = -p^2 + 3pq - 3q^2$
2. Required Sum: $(4pq - p^2 - 3q^2) + (2p^2 + 6q^2) + (-p^2 + 3pq - 3q^2)$
$= 7pq$

Solutions to Chapter 4

Question 4.1: Find the value of:

1. $5x \times 7 = 35x$

2. $2b \times 3 = 6b$

3. $x^2 \times x^3 = x^5$

4. $5x \times 6x^2 = 30x^3$

5. $6c^3 \times 7c^4 = 42c^7$

6. $9y^2 \times 5y^5 = 45y^7$

7. $3m^2 \times 5m^5 = 15m^7$

8. $4a^4 \times 6a^6 = 24a^{10}$

9. $3x \times 4y = 12xy$

10. $5a \times 6b^2 = 30ab^2$

11. $4c^2 \times 5d^5 = 20c^2d^5$

12. $3p^4 \times 5q^5 = 15p^4q^5$

13. $6ax \times 5ax = 30a^2x^2$

14. $3qr \times 4qr = 12q^2r^2$

15. $ab \times ab = a^2b^2$

16. $3ac \times 5ad = 15a^2cd$

17. $a^3x \times a^4x^3 = a^7x^3$

18. $3x^3y^2 \times 4y^5 = 12x^3y^7$

19. $a^3b^5 \times a^5b^4 = a^8b^9$

20. $a^4 \times 3a^5b^2 = 3a^9b^2$

Question 4.2: Multiply

1. $ab - ac$ by $a^2c = a^3bc - a^3c$
2. $x^2y - x^2z + 4yz^5$ by $x^3yz^3 = x^5y^2z^3 - x^5yz^4 + 4x^3y^2z^8$
3. $5a^2 - 3b^2$ by $3ab^2c^4 = 15a^3b^2c^4 - 9ab^4c^4$
4. $a^2b - 5ab + 6a$ by $3a^3b = 3a^5b^2 - 15a^4b^2 + 18a^4b$
5. $a^2 - 2b^3$ by $3x^2 = 3a^2x^2 - 6b^3x^2$
6. $2ax^2 - b^2y + 3$ by $a^2xy = 2a^3x^3y - a^2b^2xy^2 + 3a^2xy$
7. $7p^2q - pq^2 + 1$ by $2p^2 = 14p^4q - 2p^3q^2 + 2p^2$
8. $m^2 + 5mn - 3n^2$ by $4m^2n = 4m^4n + 20m^3n^2 - 12m^2n^3$
9. $xy^2 - 3x^2z - 2$ by $3yz = 3xy^3z - 9x^2yz^2 - 6yz$
10. $a^3 - 3a^2x$ by $2a^2bx = 2a^5bx - 6a^4bx^2$

Question 4.3: Multiply together

1. $a \times -2 = -2a$
2. $-3 \times 4x = -12x$
3. $-x^2 \times -x^3 = x^5$
4. $-5m \times 3m^3 = -15m^4$
5. $-4q \times 3q^2 = -12q^3$
6. $-4y^2 \times -4y^3 = 16y^5$
7. $-3m^3 \times 3m^3 = -9m^6$

8. $4x^4 \times -4x^4 = -16x^8$

9. $-3x \times -4y = 12xy$

10. $-5a^2 \times 4x = -20a^2x$

11. $-3p^2 \times -4q^5 = 12p^2q^5$

12. $3ab \times -4ab = -12a^2b^2$

13. $3a^2 \times -b^2 \times 2ab = -6a^3b^3$

14. $-a \times -b \times -c^2 = -abc^2$

15. $3a^2 \times -2b \times -4c^3 \times -d = -12a^2bc^3d$

16. $6a^2 - 5a^2b - 4ab^2 \times -3ab^2 = -18a^3b^2 + 15a^3b^3 + 12a^2b^4$

17. $-ab + ac - bc \times -ab = a^2b^2 - a^2bc + ab^2c$

18. $a^2c - ac^2 + c^4 \times -a^3c = -a^5c^2 + a^4c^3 - a^3c^5$

19. $-3a^2 - 4ax + 5x^2 \times -a^2x^3 = 3a^4x^3 + 4a^3x^4 - 5a^2x^5$

20. $-2ab + cd - ef \times -3x^2y^2 = 6abx^2y^2 - 3cdx^2y^2 + 3efx^2y^2$

Question 4.4: Find the product of:

1. $a + 7 \times a + 5 = a^2 + 12a + 35$

2. $x - 3 \times x + 4 = x^2 + x - 12$

3. $a - 6 \times a - 7 = a^2 - 13a + 42$

4. $y - 4 \times y + 4 = y^2 - 16$

5. $x + 9 \times x - 8 = x^2 + x - 72$

6. $c - 8 \times c + 8 = c^2 - 64$

7. $k + 5 \times k - 5 = k^2 - 25$

8. $m - 9 \times m + 12 = m^2 + 3m - 108$

9. $x - 12 \times x + 11 = x^2 - x - 132$

10. $a - 14 \times a + 1 = a^2 - 13a - 14$

11. $p - 10 \times p + 10 = p^2 - 100$

12. $d + 7 \times d + 7 = d^2 + 28d + 49$

13. $x - 4 \times -x + 4 = 16 - x^2$

14. $-y + 3 \times -y - 3 = y^2 - 9$

15. $-a + 4 \times -a + 5 = a^2 - 9a + 20$

16. $3y - 5 \times y + 7 = 3y^2 + 16y - 35$

17. $5m - 4 \times 7m - 3 = 35m^2 - 43m + 12$

18. $7b + c \times 7b - 2c = 49b^2 - 7bc - 2c^2$

19. $xy + 2b \times xy - 2b = x^2y^2 - 4b^2$

20. $3x - 4y \times 2a + 3b = 6ax + 9bx - 8ay - 12by$

Question 4.5: Multiply together

1. $x^2 - 3x - 2$ by $2x - 1 = 2x^3 - 7x^2 - x + 2$

2. $4a^2 - a - 2$ by $2a + 3 = 8a^3 + 10a^2 - 7a - 6$

3. $2y^2 - 3y + 1$ by $3y - 1 = 6y^3 - 11y^2 + 6y - 1$

4. $3x^2 + 4x + 5$ by $4x - 5 = 12x^3 + x^2 - 25$

5. $2a^2 - 3a - 6$ by $a - 2 = 2a^3 - 7a^2 + 12$

6. $5b^2 - 2b + 3$ by $-2b - 3 = -10b^3 - 11b^2 + 9$

7. $3x^2 - 2x + 7$ by $2x - 7 = 6x^3 - 25x^2 - 49$

8. $5c^2 - 4c + 3$ by $-2c + 1 = -10c^3 + 13c^2 - 10c + 3$

9. $x^2 + x - 2$ by $x^2 + x - 2 = x^4 + 2x^3 - 3x^2 - 4x + 4$

10. $x^2 + x - 2$ by $x^2 - x + 2 = x^4 - x^2 + 4x - 4$

11. $a + 3$ by $a - 2 = a^2 + a - 6$

12. $a - 7$ by $a - 6 = a^2 - 13a + 42$

13. $x - 4$ by $x + 5 = x^2 + x - 20$

14. $b - 6$ by $b + 4 = b^2 + 2b - 24$

15. $y - 7$ by $y - 1 = y^2 - 8y + 7$

16. $a - 1$ by $a - 9 = a^2 - 10a + 9$

17. $c - 5$ by $c + 4 = c^2 - c - 20$

18. $x - 9$ by $x - 3 = x^2 - 12x + 27$

19. $y - 4$ by $y + 7 = y^2 + 3y - 28$

20. $a - 3$ by $a + 9 = a^2 + 6a - 27$

Solutions to Chapter 5

Question 5.1: Divide:

1. $\dfrac{2x^3}{x^3} = 2$

2. $\dfrac{6a^5}{3a} = 2a^4$

3. $\dfrac{5a^7}{a^4} = 5a^3$

4. $\dfrac{21b^7}{7b^3} = 3b^4$

5. $\dfrac{x^3 y^2}{-xy} = -x^2 y$

6. $\dfrac{-3xy^3}{3y} = -xy^2$

7. $\dfrac{4p^2 q^3}{-2pq} = -2pq^2$

8. $\dfrac{15m^3n}{-5mn} = -3m^2$

9. $\dfrac{-l^3m^2}{-lm} = l^2m$

10. $\dfrac{-48x^9}{-6x^3} = 8x^6$

11. $\dfrac{35z^{11}}{-7z^5} = -5z^6$

12. $\dfrac{-7a^3b}{-7b} = a^3$

13. $\dfrac{-28p^5q}{28p^5} = -q$

14. $\dfrac{24xyz^3}{-3z^2} = -8xyz$

15. $\dfrac{-12b^2c^5}{6b^2c^5} = -2$

16. $\dfrac{-9k^{11}}{-k^{11}} = 9$

17. $\dfrac{2k^3l^5}{-kl} = -2k^2l^4$

18. $\dfrac{-45a^4b^3c^{15}}{9a^2b^3c^{10}} = 5a^2c^5$

19. $\dfrac{-186a^2b^2cx^2}{-7abx^2} = 186abc$

20. $\dfrac{5a^3b - 7ab^3}{ab} = 5a^2 - 7b^2$

21. $\dfrac{3x^2 - 2x}{x} = 3x - 2$

22. $\dfrac{x^2 - xy - xz}{-x} = -x + y + z$

23. $\dfrac{10a^3 - 5a^2b + a}{-a} = -10a^2 + 5ab - 1$

24. $\dfrac{4x^3 + 36ax^2 - 16x}{-4x} = -x^2 - 9ax + 4$

25. $\dfrac{3a^3 - 9a^2b - 6ab^2}{-3a} = -a^2 + 3ab + 3b^2$

Question 5.2: Divide:

1. $\dfrac{a^2 + 2a + 1}{a + 1} = a + 1$

2. $\dfrac{b^2 + 3b + 2}{b + 2} = b + 1$

3. $\dfrac{x^2 + 4x + 3}{x + 1} = x + 3$

4. $\dfrac{y^2 + 5y + 6}{y + 3} = y + 2$

5. $\dfrac{x^2 + 5x - 6}{x - 1} = x + 6$

6. $\dfrac{x^2 + 2x - 8}{x - 2} = x + 4$

7. $\dfrac{p^2 + 3p - 40}{p + 8} = p - 5$

8. $\dfrac{q^2 - 4q - 32}{q + 4} = q - 8$

9. $\dfrac{a^2 + 5a - 50}{a + 10} = a - 5$

10. $\dfrac{m^2 + 7m - 78}{m - 6} = m + 13$

11. $\dfrac{x^2 + ax - 30a^2}{x + 6a} = x - 5a$

12. $\dfrac{a^2 + 9ab - 36b2}{a + 12b} = a - 3b$

13. $\dfrac{2x^2 - 13x - 24}{2x + 3} = x - 8$

14. $\dfrac{5x^2 + 16x + 3}{x + 3} = 5x + 1$

15. $\dfrac{6x^2 + 5x - 21}{2x - 3} = 3x + 7$

16. $\dfrac{12a^2 + ax - 6x^2}{3a - 2x} = 4a + 3x$

17. $\dfrac{-5x^2 + xy + 6y^2}{-x - y} = 5x - 6y$

18. $\dfrac{6a^2 - ac - 35c^2}{2a - 5c} = 3a + 7c$

19. $\dfrac{12p^2 - 74pq + 12q^2}{2p - 12q} = 6p - q$

20. $\dfrac{12^2 - 31ab + 20b^2}{4a - 5b} = 3a - 4b$

Solutions to Chapter 6

Question 6.1: Simplify by removing brackets

1. $3(x - 2y) - 2(x - 4y)$
   $= 3x - 6y - 2x + 8y = x + 2y$

2. $x - 3(y - x) - 4(x - 2y)$
   $= x - 3y + 3x - 4x + 8y = 5y$

3. $16 - 3(2x - 3) - (2x + 3)$
   $= 16 - 6x + 9 - 2x - 3 = 22 - 8x$

4. $4(x + 3) - 2(7 + x) + 2$
   $= 4x + 12 - 14 - 2x + 2 = 2x$

5. $8(x - 3) - (6 - 2x) - 2(x + 2) + 5(5 - x)$
   $3x - 9$

99

6. $2x - 5(3x - 7 + y) + 4(2x + 3y - 8) - 7y$
   $-5x + 3$

7. $2x - 5(3x - 7(4x - 9))$
   $= 2x - 5(-25x + 63) = 127x - 315$

8. $x^3 + 3(x^2y + xy^2) + y^3 - x^3 - 3(x^2y - xy^2) - y^3$
   $= x^3 + 3x^2y + 3xy^2 + y^3 - x^3 - 3x^2y + 3xy^2 - y^3 = 6xy^2$

9. $4x - 3(x - (1 - y) + 2(1 - x))$
   $= 4x - 3x + 3(1 - y) - 6(1 - x)$
   $= 7x - 3y - 3$

10. $x - (y - z)(x - y - z - 2(y + z))$
    $= x - (y - z)(x - 3y - 3z)$
    $= x - x(y - z) + 3y(y - z) + 3z(y - z)$
    $= x - xy + xz + 3y^2 - 3yz + 3yz - 3z^2$
    $= x - xy + xz + 3y^2 - 3z^2$

Question 6.2: Find the sum of $a - 2b + c$, $3b - (a - c)$, $(3a - b) + 3c$.

$a - 2b + c + 3b - (a - c) + (3a - b) + 3c$
$= a - 2b + c + 3b - a + c + 3a - b + 3c = 3a + 5c$

Question 6.3: Subtract $1 - x^2$ from 1, and add the result to $2y - x^2$

$1 - (1 - x^2) + (2y - x^2) = 1 - 1 + x^2 + 2y - x^2 = 2y$

Question 6.4: Simplify $a + 2b - 3c + (b - 3a + 2c) - (3b - 2a - 2c)$.

$a + 2b - 3c + (b - 3a + 2c) - (3b - 2a - 2c)$
$= a + 2b - 3c + b - 3a + 2c - 3b + 2a + 2c$
$= c$

Question 6.5: Find the product of $3x^2y, 2xy^2, -7x^3, -5x^4y^5$.

$3x^2y \times 2xy^2 \times -7x^3 \times -5x^4y^5$
$= 210x^{10}y^8$

Question 6.6: Simplify $2x^2 - (2xy - 3y^2) + 4y^2 + (5xy - 2x^2) + x^2 - (2xy + 6y^2)$.

$2x^2 - (2xy - 3y^2) + 4y^2 + (5xy - 2x^2) + x^2 - (2xy + 6y^2)$
$= 2x^2 - 2xy + 3y^2 + 4y^2 + 5xy - 2x^2 + x^2 - 2xy - 6y^2$
$= x^2 + y^2 + xy$

Solutions to Chapter 7

Question 7.1: Write down the solution to the following equations:

1. $7x = 21$, therefore $x = 3$

2. $3x = 15$, therefore $x = 5$

3. $9x = 18$, therefore $x = 2$

4. $5x = 5$, therefore $x = 3$

5. $12x = 132$, therefore $x = 11$

6. $33 = 11x$, therefore $x = 3$

7. $4x = -12$, therefore $x = -3$

8. $-10 = -5x$, therefore $x = 2$

9. $4x = 18$, therefore $x = 9/2$

10. $12x = 42$, therefore $x = 7/2$

11. $30 = -6x$, therefore $x = -5$

12. $4x = 0$, therefore $x = 0$

13. $6x = 26$, therefore $x = 13/3$

14. $0 = 11x$, therefore $x = 0$

15. $1 = 11x$, therefore $x = 1/11$

16. $3x = -27$, therefore $x = -9$

17. $0 = -2x$, therefore $x = 0$

18. $6x = 3$, therefore $x = 1/2$

19. $5 = 15x$, therefore $x = 1/3$

20. $-24 = -8x$, therefore $x = 3$

Elementary Algebra

Question 7.2: Solve the following equations:

1. $6x + 3 = 15$, therefore $x = (15 - 3)/6 = 2$

2. $5x - 7 = 28$, therefore $x = (28 + 7)/5 = 7$

3. $13 = 7 + 2x$, therefore $x = (13 - 7)/2 = 3$

4. $15 = 37 - 11x$, therefore $x = (37 - 15)/11 = 2$

5. $4x - 7 = 11$, therefore $x = (11 + 7)/4 = 9/2$

6. $7x = 18 - 2x$, therefore $x = 18/9 = 2$

7. $3x - 18 = 7 - 2x$, therefore $x = (7 + 18)/(3 + 2) = 5$

8. $4x = 13 - 2x - 10$, therefore $x = 3/6 = 1/2$

9. $3x = 7 - 2x + 8$, therefore, $x = 15/5 = 3$

10. $5x - 17 + 3x - 5 = 6x - 7 - 8x + 115$, $x = (108 + 22)/10 = 13$

11. $5(x - 3) = 4(x - 2)$, therefore $x = (15 - 8)/(5 - 4) = 7$

12. $11(5 - 4x) = 7(5 - 6x)$, therefore $x = (55 - 35)/2 = 10$

13. $3 - 7(x + 1) = 5 - 4x$, therefore $x = (5 + 4)/(-7 + 4) = -3$

14. $5 - 4(x - 3) = x - 2(x - 1)$, therefore $x = (17 - 2)/3 = 5$

15. $8(x - 3) - 2(3 - x) = 2(x + 2) - 5(5 - x)$
    Therefore $x = (4 - 25 + 30)/3 = 3$

16. $(x + 2)(x + 3) + (x - 3)(x - 2) - 2x(x + 1) = 0$
    $x^2 + 5x + 6 + x^2 - 5x + 6 - 2x^2 - 2x = 0$, therefore $x = 6$

17. $(2x + 1)(2x + 6) - 7(x - 2) = 4(x + 1)(x - 1) - 9x$
    $4x^2 + 14x + 6 - 7x + 14 = 4x^2 - 4 - 9x$
    Therefore $x = (-4 - 20)/16 = -3/2$

18. $(3x + 1)^2 + 6 + 18(x + 1)^2 = 9x(3x - 2) + 65$
    $9x^2 + 6x + 1 + 6 + 18x^2 + 36x + 18 = 27x^2 - 18x + 65$
    $42x + 25 = -18x + 65$
    Therefore $60x = 40 \implies x = 2/3$

19. Show that $x = 5$ satisfies $5x - 6(x - 4) = 2(x + 5) + 5(x - 4) - 6$
    $5x - 6(x - 4) = 2(x + 5) + 5(x - 4) - 6$
    $5x - 6x + 24 = 2x + 10 + 5x - 20 - 6$
    Therefore $x = (10 - 20 - 24 - 6)/(-1 - 2 - 5) = 5$
    Therefore $x = 5$ satisfies the above equation.

20. Verify that $x = 3$ satisfies $2(x + 1)(x + 3) + 8 = (2x + 1)(x + 5)$
    $2(x + 1)(x + 3) + 8 = (2x + 1)(x + 5)$
    $2x^2 + 8x + 6 = 2x^2 + 11x + 5$
    Therefore $x = 3$.
    Therefore $x = 3$ satisfies the above equation.

Solutions to Chapter 8

Express the solutions to the following problems as symbolic expressions.

Question 8.1: By how much does x exceed 5?
$x - 5$

Question 8.2: By how much is y less than 15?
$15 - y$

Question 8.3: What must be added to a to make 7?
$7 - a$

Question 8.4: By what must 5 be multiplied to make a?
$a/5$

Question 8.5: What is the quotient when 3 is divided by a?
$3/a$

Question 8.6: By what must $6x$ be divided to get 2?
$6x/2 = 3x$

Question 8.7: By how much does $6x$ exceed $2x$?

$6x - 2x = 4x$

Question 8.8: The sum of two numbers is x and one of the numbers is 10. What is the other number?

$x - 10$

Question 8.9: The sum of three numbers is 100 and one of the numbers is 25, other is x, what is the third number?

$100 - 25 - x = 75 - x$

Question 8.10: The product of two numbers is $4x$. One of the factors is 4. What is the other factor?

x

Question 8.11: The product of two number p and one of them is m. What is the other number?

p/m

Question 8.12: How many times is x contained in $2y$?

$2y/x$ times.

Question 8.13: The difference between two numbers is 8, and the greater of them is a; what is the other?

$a - 8$

Question 8.14: The difference between two numbers is x; and the lesser of them is 6; what is the other?

$x + 6$

Question 8.15: What number is 30 less than y?

$y - 30$

Question 8.16: The sum of 12 equal numbers is $48x$. What is the value of each number?

$4x$

Question 8.17: How many numbers, each of value y must be taken to make $15xy$?

$15x$

Question 8.18: If there are x numbers each equal to $2a$; what is their sum?

$2ax$

Question 8.19: If there are 5 numbers each equal to x; what is their product?

x^5

Question 8.20: If there are x numbers each equal to p, what is their product?

x^p

Question 8.21: If there are n books each worth Rs. y; what is the total cost of the books?

Rs. ny

Question 8.22: How many books each costing Rs. 2 be bought for Rs. y?

$y/2$ books.

Question 8.23: What is the cost of x apples at Rs y a dozen?

Rs. $xy/12$

Question 8.24: What is the cost of n oranges at Rs m a score?

Rs. $mn/20$

Question 8.25: How many hours will it take to travel x miles at y miles an hour?

x/y hours

Question 8.26: How far can I walk in p hours at the rate of q miles per hour?

pq miles

Question 8.27: If I can walk m miles in n days, how many miles am I walking every day?

m/n miles

Question 8.28: In 5 years from now, my age will be half of that of my father. How old is my father now, if my age is x years today?

$2(x+5) - 5 = 2x + 5$ years

Question 8.29: The sum of 3 consecutive numbers is 27. What are the numbers?

8, 9 and 10.

Question 8.30: The number of 2 consecutive odd numbers is always divisible by 4. Is this true?

Two consecutive odd numbers are $2n - 1$ and $2n + 1$. Their sum is $4n$, which is always divisible by 4.

Question 8.31: The sum of three consecutive odd numbers is 33. What are the numbers?

9, 11 and 13

Question 8.32: What is the next odd number after $2n - 1$?

$2n + 1$

Question 8.33: How old is a man who in x years will be n times as old as his son's age today, given that his son is y years old today?

$n(x + y) - x$ years

Question 8.34: How old were you x years ago, given that you will be y years old, z years from now?

Today's age is p. Hence, $p + z = y \implies p = y - z$ Age x years ago is
$p - x = y - z - x$ years

Question 8.35: What is the cost of $6x$ plums and $4x$ nuts; given that a nut costs c and m plums are y times the cost of z nuts?

Nut cost is c, plum cost is p

$$mp = yzc \implies p = \frac{yzc}{m}$$

$$\text{Cost} = 6xp + 4xc = 6x\frac{yzc}{m} + 4xc = (2m + 3zy)\frac{2xc}{m}$$

Solutions to Chapter 9

Question 9.1: Find the highest common factor of:

1. $3ab^2, 2ab^3$
   HCF of $3ab^2, 2ab^3$ is ab^2

2. $x^3y^2, 4x^2y^5$
   HCF of $x^3y^2, 4x^2y^5$ is x^2y^2

3. $2x^3y^2, 4x^4y^5$
   HCF of $2x^3y^2, 4x^4y^5$ is $2x^3y^2$

4. $4x^5, 2xy^2z^3$
   HCF of $4x^5, 2xy^2z^3$ is $2x$

5. a^2b^2c, a^3bc^3
   HCF of a^2b^2c, a^3bc^3 is a^2bc

6. $3a^2b, 9abc$
   HCF of $3a^2b, 9abc$ is $3ab$

7. $6x^2y^2z, 2xy$
   HCF of $6x^2y^2z, 2xy$ is $2xy$

8. $15y^3, 5xy^6z^2$
   HCF of $15y^3, 5xy^6z^2$ is $5y^3$

9. $12a^3bc^2, 18ab^2c^3$
   HCF of $12a^3bc^2, 18ab^2c^3$ is $6abc^2$

10. $7x^3y^5z^4, 21x^2yz^3$
    HCF of $7x^3y^5z^4, 21x^2yz^3$ is $7x^2yz^3$

Question 9.2: Find the lowest common multiple of:

1. $7x^3y^5z^4, 21x^2yz^3$
   LCM of $7x^3y^5z^4, 21x^2yz^3$ is $21x^3y^5z^4$

2. a^2b^4, abc
   LCM of a^2b^4, abc is a^2b^4c

3. $2x^3y, 3xy^2z$
   LCM of $2x^3y, 3xy^2z$ is $6x^3y^2z$

4. $4a^2, 3abx^4$
   LCM of $4a^2, 3abx^4$ is $12a^2bx^4$

5. $4a^4bc^3, 5ab^2$
   LCM of $4a^4bc^3, 5ab^2$ is $20a^4b^2c^3$

6. $2ab, 4xy$
   LCM of $2ab, 4xy$ is $4abxy$

7. mn, nl, lm
   LCM of mn, nl, lm is lmn

8. $xy^2, 3yz^2, 2zx^3$
   LCM of $xy^2, 3yz^2, 2zx^3$ is $6x^3y^2z^2$

9. $2xy, 3yz, 4zx$
   LCM of $2xy, 3yz, 4zx$ is $12xyz$

10. p^2qr, pq^2r, pqr^2
    LCM of p^2qr, pq^2r, pqr^2 is $p^2q^2r^2$

Question 9.3: Reduce to lowest terms

1. $\dfrac{2a}{4ab}$

   $\dfrac{2a}{4ab} = \dfrac{1}{2b}$

2. $\dfrac{3a^2}{9ab}$

   $\dfrac{3a^2}{9ab} = \dfrac{a}{3b}$

3. $\dfrac{2bc^2}{6b^2c}$

   $\dfrac{2bc^2}{6b^2c} = \dfrac{c}{3b}$

4. $\dfrac{2abc}{8a^2bc^2}$

   $\dfrac{2abc}{8a^2bc^2} = \dfrac{1}{4ac}$

5. $\dfrac{xy^2z^3}{x^3y^4z}$

   $\dfrac{xy^2z^3}{x^3y^4z} = \dfrac{z^2}{x^2y^2}$

6. $\dfrac{12mn}{15lm}$

   $\dfrac{12mn}{15lm} = \dfrac{4n}{5l}$

7. $\dfrac{14xy^3}{21x^3z^2}$

   $\dfrac{14xy^3}{21x^3z^2} = \dfrac{2y^3}{x^2z^2}$

8. $\dfrac{9a^3b}{12ab^3c}$

   $\dfrac{9a^3b}{12ab^3c} = \dfrac{3a^2}{4b^2c}$

9. $\dfrac{15a^2b^2c^3}{18abc^2}$

   $\dfrac{15a^2b^2c^3}{18abc^2} = \dfrac{5abc}{6}$

10. $\dfrac{5a^3y^2z^4}{15ay^4z}$

$\dfrac{5a^3y^2z^4}{15ay^4z} = \dfrac{a^2z^3}{3y^2}$

Question 9.4: Simplify the following expressions:

1. $\dfrac{xy}{ab} \times \dfrac{a^2b^3}{xy^2}$

$\dfrac{xy}{ab} \times \dfrac{a^2b^3}{xy^2} = \dfrac{ab^2}{y}$

2. $\dfrac{ab}{2cd^3} \times \dfrac{4c^2d}{ab^3}$

$\dfrac{ab}{2cd^3} \times \dfrac{4c^2d}{ab^3} = \dfrac{2c}{b^2d^2}$

3. $\dfrac{2ax^2}{3y^3z} \times \dfrac{yz^3}{4a^2x}$

$\dfrac{2ax^2}{3y^3z} \times \dfrac{yz^3}{4a^2x} = \dfrac{xz^2}{6ay^2}$

4. $\dfrac{6a^2x^3}{7ab^2} \times \dfrac{14b^2c}{12ax}$

$\dfrac{6a^2x^3}{7ab^2} \times \dfrac{14b^2c}{12ax} = \dfrac{cx^2}{1}$

5. $\dfrac{3ab^2}{5b^3c} \times \dfrac{15b^2c^2}{9a^2b}$

$\dfrac{3ab^2}{5b^3c} \times \dfrac{15b^2c^2}{9a^2b} = \dfrac{c}{a}$

6. $\dfrac{3p^2q^2}{9xy} \Big/ \dfrac{pq}{x^2y^2}$

$\dfrac{3p^2q^2}{9xy} \Big/ \dfrac{pq}{x^2y^2} = \dfrac{3p^2q^2}{9xy} \times \dfrac{x^2y^2}{pq} = \dfrac{pqxy}{3}$

7. $\dfrac{10b^2}{4x^2} \Big/ \dfrac{b^2p^2}{3x^6}$

$\dfrac{10b^2}{4x^2} \Big/ \dfrac{b^2p^2}{3x^6} = \dfrac{10b^2}{4x^2} \times \dfrac{3x^6}{b^2p^2} = \dfrac{15x^4}{2p^2}$

8. $\dfrac{17y}{x^2z^3}\Big/\dfrac{34y^3}{x^5z}$

$\dfrac{17y}{x^2z^3}\Big/\dfrac{34y^3}{x^5z} = \dfrac{17y}{x^2z^3} \times \dfrac{x^5z}{34y^3} = \dfrac{x^3}{2y^2z^2}$

9. $\dfrac{9ax^2}{5a^2z}\Big/\dfrac{x^3y^2}{2a^2y}$

$\dfrac{9ax^2}{5a^2z}\Big/\dfrac{x^3y^2}{2a^2y} = \dfrac{9ax^2}{5a^2z} \times \dfrac{2a^2y}{x^3y^2} = \dfrac{18a}{5xyz}$

10. $\dfrac{14d^3}{abc}\Big/\dfrac{81d^3}{27a^2b^2c^2}$

$\dfrac{14d^3}{abc}\Big/\dfrac{81d^3}{27a^2b^2c^2} = \dfrac{14d^3}{abc} \times \dfrac{27a^2b^2c^2}{81d^3} = \dfrac{abc}{3}$

Question 9.5: Simplify the following expressions:

1. $\dfrac{a}{2} + \dfrac{a}{3}$
LCM of denominators 2 and 3 is 6
$\dfrac{a}{2} + \dfrac{a}{3} = \dfrac{3a}{6} + \dfrac{2a}{6}$
$= \dfrac{5a}{6}$

2. $\dfrac{b}{3} + \dfrac{b}{4}$
LCM of denominators 3 and 4 is 12
$\dfrac{b}{3} + \dfrac{b}{4} = \dfrac{4b}{12} + \dfrac{3b}{12}$
$= \dfrac{7b}{12}$

3. $\dfrac{x}{4} - \dfrac{x}{5}$
LCM of denominators 4 and 5 is 20
$\dfrac{x}{4} - \dfrac{x}{5} = \dfrac{5x}{20} + \dfrac{4x}{20}$
$= \dfrac{9x}{20}$

4. $\dfrac{2y}{3} + \dfrac{y}{6}$
LCM of denominators 3 and 6 is 6

$$\frac{2y}{3} + \frac{y}{6} = \frac{4y}{6} + \frac{y}{6}$$

$$= \frac{5y}{6}$$

5. $\dfrac{a}{5} - \dfrac{a}{6}$

LCM of denominators 5 and 6 is 30

$$\frac{a}{5} - \frac{a}{6} = \frac{6a}{30} - \frac{5a}{30}$$

$$= \frac{a}{30}$$

6. $\dfrac{m}{8} - \dfrac{2n}{20}$

LCM of denominators 8 and 20 is 40

$$\frac{m}{8} - \frac{2n}{20} = \frac{5m}{40} - \frac{4n}{40}$$

$$= \frac{n}{40}$$

7. $\dfrac{2a}{3} + \dfrac{4a}{9b}$

LCM of denominators 3 and $9b$ is $9b$

$$\frac{2a}{3} + \frac{4a}{9b} = \frac{6ab}{9b} + \frac{4a}{9b}$$

$$= \frac{2a(3b + 2)}{9b}$$

8. $\dfrac{ab}{3} - \dfrac{x^2y}{6xy}$

LCM of denominators 3 and $6xy$ is $6xy$

$$\frac{ab}{3} - \frac{x^2y}{6xy} = \frac{2abxy}{6xy} - \frac{x^2y}{6xy}$$

$$= \frac{xy(2ab - x)}{6xy} = \frac{2ab - x}{6}$$

9. $\dfrac{a}{xy} + \dfrac{2a}{yz} - \dfrac{3a}{xz}$

LCM of denominators xy, yz and xz is xyz

$$\frac{a}{xy} + \frac{2a}{yz} - \frac{3a}{xz} = \frac{az}{xyz} + \frac{2ax}{xyz} - \frac{3ay}{xyz}$$

$$= \frac{a(z + x - 3y)}{xyz}$$

10. $\dfrac{a^3}{3a^2b} - \dfrac{a^3}{ab^2} + \dfrac{ac}{6bc}$

LCM of denominators $3a^2b, ab^2$ and $6bc$ is $6a^2b^2c$

$$\frac{a^3}{3a^2b} - \frac{a^3}{ab^2} + \frac{ac}{6bc}$$

$$= \frac{2a^3bc}{6a^2b^2c} - \frac{6a^4c}{6a^2b^2c} + \frac{a^3bc}{6a^2b^2c}$$

$$= \frac{2a^3bc - 6a^4c + a^3bc}{6a^2b^2c} = \frac{a^2c(2ab - 6a^2 + ab)}{6a^2b^2c}$$

$$= \frac{ab - 2a^2}{3b^2}$$

We have solved this problem to demonstrate an important point. Sometimes, we miss canceling out a few factors to begin with, which results in large LCM. Do not panic. Go through the steps systematically and convert the final solution to its lowest terms.

The elegant solution to this problem is shown below.

$$\frac{a^3}{3a^2b} - \frac{a^3}{ab^2} + \frac{ac}{6bc} = \frac{a}{3b} - \frac{a^2}{b^2} + \frac{a}{6b}$$

The LCM of denominators is $6b^2$

$$= \frac{2ab}{6b^2} - \frac{6a^2}{6b^2} + \frac{ab}{6b^2}$$

$$= \frac{2ab - 6a^2 + ab}{6b^2} = \frac{ab - 2a^2}{3b^2}$$

Solutions to Chapter 10

Question 10.1: Solve the following equations:

1. $x + y = 19, x - y = 7$
   $x + y = 19$
   $x - y = 7$
   Adding the above equations, we get
   $2x = 26$; therefore $x = 13$
   Substituting value of x in (1), we get
   $13 + y = 19$; therefore $y = 6$
   $x = 13$ and $y = 6$ is the required solution.

2. $x + y = 23, x - 5 = 5$
   $x + y = 23$
   $x - 5 = 5$

Adding the above equations, we get
$2x = 28$; therefore $x = 14$
Substituting value of x in (1), we get
$14 + y = 23$; therefore $y = 9$
$x = 14$ and $y = 9$ is the required solution.

3. $x + y = 11, x - y = -9$
$x + y = 11$
$x - y = -9$
Adding the above equations, we get
$2x = 2$, therefore $x = 1$
Substituting value of x in (1), we get
$1 + y = 11$, therefore $y = 10$
$x = 1$ and $y = 10$ is the required solution.

4. $x + y = 24, x - y = 0$
$x + y = 24$
$x - y = 0$
Adding the above equations, we get
$2x = 24$, therefore $x = 12$
Substituting value of x in (1), we get
$12 + y = 24$, therefore, $y = 12$
$x = 12$ and $y = 12$ is the required solution.

5. $x + y = 6, x - y = 0$
$x + y = 6$
$x - y = 0$
Adding the above equations, we get
$2x = 6$, therefore $x = 3$
Substituting value of x in (1), we get
$3 + y = 6$, therefore $y = 3$
$x = y = 3$ is the required solution.

6. $x - y = 25, x + y = 13$
$x + y = 13$
$x - y = 25$
Adding the above equations, we get
$2x = 38$, therefore $x = 19$
Substituting value of x in (1), we get
$19 + y = 13$, therefore $y = -6$
$x = 19$ and $y = -6$ is the required solution.

7. $3x + 5y = 50, 4x + 3y = 41$
$3x + 5y = 50$
$4x + 3y = 41$
First equation $\times 4$
$12x + 20y = 200$
Second equation $\times 3$
$12x + 9y = 123$
Subtracting the above equations, we get
$11y = 77$, therefore, $y = 7$
Substituting value of y in (1), we get
$3x + 35 = 50$, therefore $x = 5$
$x = 5$ and $y = 7$ is the required solution.

8. $x + 5y = 18, 3x + 2y = 41$
$x + 5y = 18$
$3x + 2y = 41$
First equation $\times 3$
$3x + 15y = 54$
Subtracting the above equations, we get
$13y = 13$, therefore $y = 1$
Substituting the value of y in the first equation. we get
$x + 5 = 18$, therefore, $x = 13$
$x = 13$ and $y = 1$ is the required solution.

9. $4x + y = 10, 5x + 7y = 47$
$4x + y = 10$
$5x + 7y = 47$
First equation $\times 7$
$28x + 7y = 70$
Subtracting the above equations, we get
$23x = 23$ or $x = 1$
Substituting value of x in (1), we get
$4 + y = 10$ or $y = 6$
$x = 1$ and $y = 6$ is the required solution.

10. $7x - 6y = 25, 5x + 4y = 51$
$7x - 6y = 25$, therefore $35x - 30y = 125$
$5x + 4y = 51$, therefore $35x + 28y = 357$
Subtracting the above equations, we get
$58y = 232$, therefore $y = 4$
Substituting value of y in (1), we get

$7x - 24 = 25$, therefore $x = 7$
$x = 7$ and $y = 4$ is the required solution.

11. $11x - 7y = 43, 2x - 3y = 13$
$11x - 7y = 43$, therefore $22x - 14y = 86$
$2x - 3y = 13$, therefore $22x - 33y = 143$
Subtracting the above equations, we get
$19y = -57$, therefore $y = -3$
Substituting value of y in (1), we get
$11x + 21 = 43$, therefore $x = 2$
$x = 2$ and $y = -3$ is the required solution.

12. $4x - 3y = 0, 7x - 4y = 36$
$4x - 3y = 0$, therefore $28x - 21y = 0$
$7x - 4y = 36$, therefore $28x - 16y = 144$
Subtracting the above equations, we get
$-5y = -144$, therefore $y = 144/5$
Substituting value of y in (1), we get
$4x - 3 \times 144/5 = 0$, therefore $x = (3 \times 144)/(4 \times 5) = 108/5$
$x = 108/5$ and $y = 144/5$ is the required solution.

13. $2x + 3y = 22, 5x + 2y = 0$
$2x + 3y = 22$, therefore $10x + 15y = 110$
$5x + 2y = 0$, therefore $10x + 4y = 0$
Subtracting the above equations, we get
$11y = 110$, therefore $y = 10$
Substituting value of y in (1), we get
$2x + 30 = 22$, therefore $x = -4$
$x = -4$ and $y = 10$ is the required solution.

14. $7x + 3y = 65, 7x - 8y = 32$
$7x + 3y = 65$
$7x - 8y = 32$
Subtracting the above equations, we get
$11y = 33$, therefore $y = 3$
Substituting value of y in (1), we get
$7x + 9 = 65$, therefore $x = 8$
$x = 8$ and $y = 3$ is the required solution.

15. $3x - 2y + z = 4, 2x + 3y - z = 3, x + y + z = 8$
$3x - 2y + z = 4$
$2x + 3y - z = 3$

$x + y + z = 8$
Adding the first two equations, we get
$5x + y = 7$
Adding the last two equations, we get
$3x + 4y = 11$
Multiply $5x + y = 7$ by 4
$20x + 4y = 28$
and take away $3x + 4y = 11$
$17x = 17$, therefore $x = 1$
Substituting the value of x in $5x + y = 7$, we get
$y = 2$
Substitute the values of x and y in the first equation, we get
$z = 5$
$x = 1, y = 2$ and $z = 5$ is 'the required solution.

16. $3x + 4y - 6z = 16, 4x + y - z = 24, x - 3y - 2z = 1$
$3x + 4y - 6z = 16$
$4x + y - z = 24$
$x - 3y - 2z = 1$
6 times $4x + y - z = 24$ minus $3x + 4y - 6z = 16$, we get
$21x + 2y = 128$
2 times $4x + y - z = 24$ minus $x - 3y - 2z = 1$, we get
$7x + 5y = 47$
3 times $7x + 5y = 47$ minus $21x + 2y = 128$, we get
$13y = 13$ or $y = 1$. Therefore $x = (128 - 2)/21 = 6$
Substituting value of x and y in (1), we get
$18 + 4 - 6z = 16$, or $z = 1$.
$x = 6, y = 1, z = 1$ is the required solution.

17. $x + 2y + 3z = 32, 4x - 5y + 6z = 27, 7x + 8y - 9z = 14$
$x + 2y + 3z = 32$
$4x - 5y + 6z = 27$
$7x + 8y - 9z = 14$
2 times $x + 2y + 3z = 32$ minus $4x - 5y + 6z = 27$, we get
$-2x + 9y = 37$
3 times $x + 2y + 3z = 32$ plus $7x + 8y - 9z = 14$, we get
$10x + 14y = 110$
5 times $-2x + 9y = 37$ plus $10x + 14y = 110$, we get
$59y = 295$ or $y = 5$
Substituting the value y in $10x + 14y = 110$, we get
$10x = 110 - 70$ or $x = 4$

Substituting value of x and y in (1), we get
$4 + 10 + 3z = 32$ or $z = 6$
$x = 4, y = 5, z = 6$ is the required solution.

18. $x - y + z = 5, 6x + 3y + 2z = 84, 3x + 4y - 5z = 13$
$x - y + z = 5$
$6x + 3y + 2z = 84$
$3x + 4y - 5z = 13$
3 times $x - y + z = 5$ plus $6x + 3y + 2z = 84$, we get
$9x + 5z = 99$
4 times $x - y + z = 5$ plus $3x + 4y - 5z = 13$, we get
$7x - z = 33$
5 times $7x - z = 33$ plus $9x + 5z = 99$, we get
$44x = 264$, or $x = 6$
Substituting the value x in $7x - z = 33$, we get
$42 - z = 33$ or $z = 9$
Substituting the value of x and z in (1) we get
$6 - y + 9 = 5$ or $y = 10$
$x = 6, y = 10, z = 9$ is the required solution.

Solutions to Chapter 11

Question 11.1: Write down the squares of each of the following expressions:

1. $a^2 b$
   $(a^2 b)^2 = a^2 b^4$

2. $3ac^3$
   $(3ac^3)^2 = 9a^2 c^6$

3. $5xy^2$
   $(5xy^2)^2 = 25x^2 y^4$

4. $6b^3 c^2$
   $(6b^3 c^2)^2 = 36b^6 c^4$

5. $4a^2 bc^3$
   $(4a^2 bc^3)^2 = 16a^4 b^2 c^6$

6. $-3x^2y^5$
   $(-3x^2y^5)^2 = 9x^4y^{10}$

7. $-2a^2b^3c$
   $(-2a^2b^3c)^2 = 4a^4b^6c^2$

8. $-3dx^4$
   $(-3dx^4)^2 = 9d^2x^8$

9. $a+b$
   $(a+b)^2 = a^2 + b^2 + 2ab$

Question 11.2: Write down the values of each of the following expressions

1. $(ab^2)^4$
   $(ab^2)^4 = a^4b^4$

2. $(-x^2y)^5$
   $(-x^2y)^5 = -x^{10}y^5$

3. $(-2m^2n^3)^6$
   $(-2m^2n^3)^6 = 64m^{12}n^{18}$

4. $(-x^3y^2)^7$
   $(-x^3y^2)^7 = -x^{21}y^{14}$

5. $(-6y^7)^4$
   $(-6y^7)^4 = 1296y^{28}$

Question 11.3: Write down the cubes of each of the following expressions

1. $2x$
   $(2x)^3 = 8x^3$

2. $3ab^2$
   $(3ab^2)^3 = 27a^3b^6$

3. $4x^3$
   $(4x^3)^3 = 64x^9$

4. $-3a^2b$
$(-3a^2b)^3 = -27a^6b^3$

5. $-4x^3y^2$
$(-4x^3y^2)^3 = -64x^9y^6$

6. $-b^2cd^3$
$(-b^2cd^3)^3 = -b^6c^3d^9$

7. $-6y^4$
$(-6y^4)^3 = -216y^{12}$

8. $-4p^3q^5$
$(-4p^3q^5)^3 = -64p^9q^{15}$

9. $a^2 - b$
$(a^2 - b)^3 = a^6 - b^3 - 3a^4b + 3a^2b^2$

Question 11.4: Write down the values of each of the following expressions

1. Square of $x + 2y$
$(x + 2y)^2 = x^2 + 4y^2 + 4xy$

2. Square of $2ab - xy$
$(2ab - xy)^2 = 4a^2b^2 + x^2y^2 + 4abxy$

3. Square of $x + y - a - b$
$(x+y-a-b)^2 = x^2+y^2+a^2+b^2+2xy-2ax-2bx-2ay-2by+2ab$

4. Cube of $x + 3y$
$(x + 3y)^3 = x^3 + 27y^3 + 9x^2y + 27xy^2$

5. Cube of $4y^2 - 3$
$(4y^2 - 3)^3 = 64y^6 - 27 + 36y^2 - 144y^4$

Solutions to Chapter 12

Question 12.1: Write down the square root of the following expressions:

1. $9x^4y^3$

   $\sqrt{9x^4y^3} = \pm 3x^2y\sqrt{y}$

2. $25a^6b^4$

   $\sqrt{25a^6b^4} = \pm 5a^3b^2$

3. $49c^2d^6$

   $\sqrt{49c^2d^6} = \pm 7cd^3$

4. $\dfrac{16x^{64}}{25}$

   $\sqrt{\dfrac{16x^{64}}{25}} = \pm \dfrac{4x^{32}}{5}$

5. $\dfrac{4x^6}{16a^4}$

   $\sqrt{\dfrac{4x^6}{16a^4}} = \pm \dfrac{2x^3}{4a^2}$

Question 12.2: Write down the cube roots of the following expressions:

1. x^6y^9

   $\sqrt[3]{x^6y^9} = x^2y^3$

2. $-a^6b^3$

   $\sqrt[3]{-a^6b^3} = -a^2b$

3. $8x^{27}$

   $\sqrt[3]{8x^{27}} = 2x^9$

4. $-27x^9$

   $\sqrt[3]{-27x^9} = -3x^3$

5. $\dfrac{-b^{27}}{27}$

$\sqrt[3]{\dfrac{-b^{27}}{27}} = -\dfrac{b^9}{3}$

Question 12.3: Write down the square root of each of the following expressions using method of inspection:

1. $a^2 - 8a + 16$
   $\sqrt{a^2 - 8a + 16} = \sqrt{(a-4)^2} = \pm(a-4)$

2. $x^2 + 14x + 49$
   $\sqrt{x^2 + 14x + 49} = \sqrt{(x+7)^2} = \pm(x+7)$

3. $64 + 48x + 9x^2$
   $64 + 48x + 9x^2 = \sqrt{(3x+8)^2} = \pm(3x+8)$

4. $25 - 30m + 9m^2$
   $\sqrt{25 - 30m + 9m^2} = \sqrt{(3m-5)^2} = \pm(3m-5)$

5. $4a^2b^4 - 12ab^2c^5 + 9c^{10}$
   $\sqrt{4a^2b^4 - 12ab^2c^5 + 9c^{10}} = \sqrt{(3c^5 - 2ab^2)^2} = \pm(3c^5 - 2ab^2)$

Solutions to Chapter 13

Question 13.1: Resolve into factors

1. $x^2 + ax$
   $x^2 + ax = x(x+a)$
   Therefore, the factors are x and $x + a$.

2. $2a^2 - 3a$
   $2a^2 - 3a = a(2a - 3)$
   Therefore, the factors are a and $2a - 3$.

3. $a^3 - a^2$
   $a^3 - a^2 = a^2(a - 1)$
   Therefore, the factors are a, a^2 and $a - 1$.

4. $a^3 - a^2b$
   $a^3 - a^2b = a^2(a - b)$
   Therefore, the factors are a, a^2 and $a - b$.

5. $3m^2 - 6mn$
   $3m^2 - 6mn = 3m(m - 2n)$
   Therefore, the factors are $3m$ and $m - 2n$.

6. $p^2 + 2p^2q$
   $p^2 + 2p^2q = p^2(1 + 2q)$
   Therefore, the factors are p, p^2 and $1 + 2q$.

7. $p^2 12p^2q$
   $p^2 - 2p^2q = p^2(1 - 2q)$
   Therefore, the factors are p, p^2 and $1 - 2q$.

8. $y^2 + xy$
   $y^2 + xy = y(y + x)$
   Therefore, the factors are y and $y + x$.

9. $12x + 48x^2y$
   $12x + 48x^2y = 12x(1 + 4xy)$
   Therefore, the factors are $12x$ and $1 + 4xy$.

10. $10c^3 - 25c^4d$
    $10c^3 - 25c^4d = 5c^3(2 - 5cd)$
    Therefore, the factors are 5, c^3 and $2 - 5cd$.

Question 13.2: Resolve into factors

1. $x^2 + xy + xz + yz$
   $x^2 + xy + xz + yz = x(x + y) + z(x + y) = (x + z)(x + y)$ Therefore, the factors are $x + y$ and $x + z$.

2. $x^2 - xz + xy - yz$
   $x^2 - xz + xy - yz = x(x - z) + y(x - z)$
   $= (x - z)(x + y)$
   Therefore, the factors are $x - z$ and $x + y$.

3. $a^2 + 2a + ab + 2b$
   $a^2 + 2a + ab + 2b = a(a + 2) + b(a + 2)$
   $= (a + 2)(a + b)$
   Therefore, the factors are $a + 2$ and $a + b$.

4. $a^2 + ac + 4a + 4c$

$a^2 + ac + 4a + 4c = a(a+c) + 4(a+c)$

$= (a+c)(a+4)$

Therefore, the factors are $a + c$ and $a + 4$.

5. $2a + 2x + ax + x^2$

$2a + 2x + ax + x^2 = 2(a+x) + x(a+x)$

$= (2+x)(a+x)$

Therefore, the factors are $2 + x$ and $a + x$.

6. $3q - 3p + pq - p^2$

$3q - 3p + pq - p^2 = 3(q-p) + p(q-p)$

$= (3+p)(q-p)$

Therefore, the factors are $p + 3$ and $q - p$.

7. $am - bm - am + bn$

$am - bm - an + bn = m(a-b) - n(a-b)$

$= (a-b)(m-n)$

Therefore, the factors are $a - b$ and $m - n$.

8. $ab - by - ay + y^2$

$ab - by - ay + y^2 = b(a-y) - y(a-y)$

$= (b-y)(a-y)$

Therefore, the factors are $b - y$ and $a - y$.

9. $pq + qr - pr - r^2$

$pq + qr - pr - r^2 = q(p+r) - r(p+r)$

$= (p+r)(q-r)$

Therefore, the factors are $p + r$ and $q - r$.

10. $2x^3 + 3 + 2x + 3x^2$

Rearrange the terms,

$2x^3 + 3 + 2x + 3x^2 = 2x^3 + 2x + 3x^2 + 3 = 2x(x^2+1) + 3(x^2+1)$

$= (2x+3)(x^2+1)$

Therefore, the factors are $2x + 3$ and $x^2 + 1$.

Question 13.3: Resolve into factors

1. $x^2 + 3x + 2$

$x^2 + 3x + 2 = x^2 + x + 2x + 2 = x(x+1) + 2(x+1)$

$= (x+2)(x+1)$

Therefore, the factors are $x + 2$ and $x + 1$.

2. $y^2 + 5y + 6$
$y^2 + 5y + 6 = y^2 + 3y + 2y + 6 = y(y + 3) + 2(y + 3)$
$= (y + 2)(y + 3)$
Therefore, the factors are $y + 2$ and $y + 3$.

3. $y^2 + 7y + 12$
$y^2 + 7y + 12 = y^2 + 4y + 3y + 12 = y(y + 4) + 3(y + 4)$
$= (y + 4)(y + 3)$
Therefore, the factors are $y + 4$ and $y + 3$.

4. $a^2 - 3a + 2$
$a^2 - 3a + 2 = a^2 - 2a - a + 2 = a(a - 2) - 1(a - 2)$
$= (a - 2)(a - 1)$
Therefore, the factors are $a - 2$ and $a - 1$.

5. $a^2 - a - 2$
$a^2 - a - 2 = a^2 - 2a + a - 2 = a(a - 2) - 1(a + 1)$
$= (a - 2)(a + 1)$
Therefore, the factors are $a - 2$ and $a + 1$.

6. $b^2 - 5b + 6$
$b^2 - 5b + 6 = b^2 - 3b - 2b + 6 = b(b - 3) - 2(b - 3)$
$= (b - 2)(b - 3)$
Therefore, the factors are $b - 2$ and $b - 3$.

7. $b^2 + 13b + 42$
$b^2 + 13b + 42 = b^2 + 7b + 6b + 42 = b(b + 7) + 6(b + 7)$
$= (b + 7)(b + 6)$
Therefore, the factors are $b + 7$ and $b + 6$.

8. $b^2 - 13b + 40$
$b^2 - 13b + 40 = b^2 - 8b - 5b + 40 = b(b - 8) - 5(b - 8)$
$= (b - 8)(b - 5)$
Therefore, the factors are $b - 8$ and $b - 5$.

9. $z^2 - 13z + 36$
$z^2 - 13z + 36 = z^2 - 9z - 4z + 36 = z(z - 9) - 4(z - 9)$
$= (z - 9)(z - 4)$
Therefore, the factors are $z - 9$ and $z - 4$.

10. $x^2 - 15x + 56$
$x^2 - 15x + 56 = x^2 - 8x - 7x + 56 = x(x - 8) - 7(x - 8)$
$= (x - 8)(x - 7)$
Therefore, the factors are $x - 8$ and $x - 7$.

Question 13.4: Resolve into factors

1. $2a^2 + 3a + 1$
   $2a^2 + 3a + 1 = 2a^2 + 2a + a + 1$
   $= 2a(a + 1) + 1(a + 1) = (2a + 1)(a + 1)$
   Therefore, the factors are $2a + 1$ and $a + 1$.

2. $3a^2 + 4a + 1$
   $3a^2 + 4a + 1 = 3a^2 + 3a + a + 1$
   $= 3a(a + 1)(1(a + 1) = (3a + 1)(a + 1)$
   Therefore, the factors are $3a + 1$ and $a + 1$.

3. $4a^2 + 5a + 1$
   $4a^2 + 5a + 1 = 4a^2 + 4a + a + 1$
   $= 4a(a + 1) + 1(a + 1) = (4a + 1)(a + 1)$
   Therefore, the factors are $4a + 1$ and $a + 1$.

4. $2a^2 + 5a + 2$
   $2a^2 + 5a + 2 = 2a^2 + 4a + a + 2$
   $= 2a(a + 2) + 1(a + 2) = (2a + 2)(a + 1)$
   Therefore, the factors are $2a + 2$ and $a + 1$.

5. $3a^2 + 10a + 3$
   $3a^2 + 10a + 3 = 3a^2 + 9a + a + 3$
   $= 3a(a + 3) + 1(a + 3) = (3a + 1)(a + 3)$
   Therefore, the factors are $3a + 1$ and $a + 3$.

6. $2a^2 + 7a + 3$
   $2a^2 + 7a + 3 = 2a^2 + 6a + a + 3$
   $= 2a(a + 3) + 1(a + 3) = (2a + 1)(a + 3)$
   Therefore, the factors are $2a + 1$ and $a + 3$.

7. $5a^2 + 7a + 2$
   $5a^2 + 7a + 2 = 5a^2 + 5a + 2a + 2$
   $= 5a(a + 1) + 2(a + 1) = (5a + 2)(a + 1)$
   Therefore, the factors are $5a + 2$ and $a + 1$.

8. $2a^2 + 9a + 10$
   $2a^2 + 9a + 10 = 2a^2 + 5a + 4a + 10$
   $= a(2a + 5) + 2(2a + 5) = (a + 2)(2a + 5)$
   Therefore, the factors are $a + 2$ and $2a + 5$.

9. $2a^2 + 7a + 6$
$2a^2 + 7a + 6 = 2a^2 + 4a + 3a + 6$
$= 2a(a+2) + 3(a+2) = (2a+3)(a+2)$
Therefore, the factors are $2a + 3$ and $a + 2$.

10. $2x^2 + 9x + 4$
$2x^2 + 9x + 4 = 2x^2 + 8x + x + 4$
$= 2x(x+4) + 1(x+4) = (2x+1)(x+4)$
Therefore, the factors are $2x + 1$ and $x + 4$.

Question 13.5: Resolve into factors

1. $a^2 - 9$
$a^2 - 9 = (a+3)(a-3)$
Therefore, the factors are $a + 3$ and $a - 3$.

2. $a^2 - 49$
$a^2 - 49 = (a+7)(a-7)$
Therefore, the factors are $a + 7$ and $a - 7$.

3. $a^2 - 81$
$a^2 - 81 = (a+9)(a-9)$
Therefore, the factors are $a + 9$ and $a - 9$.

4. $81 - 4x^2$
$81 - 4x^2 = (9 + 2x)(9 - 2x)$
Therefore, the factors are $9 + 2x$ and $9 - 2x$.

5. $a^6b^8c^4 - 9$
$a^6b^8c^4 - 9 = (a^3b^4c^2 + 3)(a^3b^4c^2 - 3)$
Therefore, the factors are $(a^3b^4c^2 + 3$ and $a^3b^4c^2 - 3$.

6. $8x^3 + 1$
$8x^3 + 1 = (2x)^3 + 1^3 = (2x+1)((2x)^2 - 2x + 1) = (2x+1)(4x^2 - 2x + 1)$
Therefore, the factors are $2x + 1$ and $4x^2 - 2x + 1$.

7. $x^3 - 8z^3$
$x^3 - 8z^3 = (x - 2z)(x^2 + 2xz + 4z^2)$
Therefore, the factors are $x - 2z$ and $x^2 + 2xz + 4z^2$.

8. $27 + x^3$
   $27 + x^3 = (3 + x)(9 - 3x + x^2)$
   Therefore, the factors are $3 + x$ and $9 - 3x + x^2$.

9. $512a^3 - 1$
   $512a^3 - 1 = (8a - 1)(64a^2 + 8a + 1)$
   Therefore, the factors are $8a - 1$ and $64a^2 + 8a + 1$.

10. $x^6 - 27z^3$
    $x^6 - 27z^3 = (x^2)^3 - (3z)^3 = (x^2 - 3z)((x^2)^2 + 3x^2z + (3z)^2)$
    $= (x^2 - 3z)(x^4 + 3x^2z + 9z^2)$
    Therefore, the factors are $x^2 - 3z$ and $x^4 + 3x^2z + 9z^2$.

Solutions to Chapter 14

Question 14.1: Solve the following equations:

1. $7(x^2 - 7) = 6x^2$
   $\implies x^2 = 49$
   $\implies x = \pm 7$

2. $(x + 8)(x - 8) = 0$
   $(x + 8)(x - 8) = 0$
   $\implies x^2 - 64 = 0$ or $x^2 = 64$
   $\implies x = \pm 8$

3. $(7 + x)(7 - x) = 0$
   $\implies 49 - x^2 = 0$
   $\implies x = \pm 7$

4. $\dfrac{x^2 + 8}{x^2 + 20} = \dfrac{1}{2}$
   $\implies 2(x^2 + 8) = (x^2 + 20)$
   $\implies x^2 = 4$
   $\implies x = \pm 2$

5. $\dfrac{11}{3 - x} = 4(x + 3)$
   $\implies 11 = 4(x + 3)(3 - x)$
   $\implies 11 = 4(9 - x^2)$
   $\implies x^2 = 25$
   $\implies x = \pm 5$

6. $\dfrac{x(3x+5)+21}{(3x-2)(2x+3)} = 1$

$\implies x(3x+5)+21 = (3x-2)(2x+3)$
$\implies 3x^2 + 5x + 21 = 6x^2 + 5x - 6$
$\implies 3x^2 = 27$
$\implies x^2 = 9$
$\implies x = \pm 3$

7. $x^2 + 2x = 8$

$\implies x2 + 2x + 1 = 9$
$\implies (x+1)^2 = 9$
$\implies (x+1) = \pm 3$
$\implies x = 2, -4$

8. $x^2 + 6x = 40$

$\implies x^2 + 6x + 9 = 49$
$\implies (x+3)^2 = 49$
$\implies (x+3) = \pm 7$
$\implies x = 4, -11$

9. $x^2 + 35 = 12x$

$\implies x^2 - 12x + 36 = 1$
$\implies (x-6)^2 = 1$
$\implies (x-6) = \pm 1$
$\implies x = 7, 5$

10. $x^2 + 15x - 34 = 0$

$\implies x^2 + 15x = 34$
$\implies x^2 + 15x + 225/4 = 34 + 225/4$
$\implies (x+15/2)^2 = 361/4$
$\implies (x+15/2) = \pm\sqrt{361/4} = \pm 19/2$
$\implies x = 2, -17$

Question 14.2: Solve the following equations:

1. $3x^2 + 2x = 21$

$x = \dfrac{-2 \pm \sqrt{2^2 - (4 \times 3 \times -21)}}{2 \times 3}$
$\implies x = \dfrac{-2 \pm 16}{6}$

2. $5x^2 = 8x + 21$

$$\implies x = \frac{--8 \pm \sqrt{(-8)^2 - (4 \times 5 \times -21)}}{2 \times 5}$$

$$\implies x = \frac{8 \pm 22}{10}$$

3. $6x^2 - x - 1 = 0$

$$\implies x = \frac{--1 \pm \sqrt{(-1)^2 - (4 \times 6 \times -1)}}{2 \times 6}$$

$$\implies x = \frac{1 \pm 5}{12}$$

4. $3 - 11x = 4x^2$

$$\implies x = \frac{-11 \pm \sqrt{11^2 - (4 \times 4 \times -3)}}{2 \times 4}$$

$$\implies x = \frac{-11 \pm 13}{8}$$

5. $21x^2 = 2x + 3$

$$\implies x = \frac{--2 \pm \sqrt{(-2)^2 - (4 \times 21 \times -3)}}{2 \times 21}$$

$$\implies x = \frac{2 \pm 16}{42}$$

6. $10 + 23x + 12x^2 = 0$

$$\implies x = \frac{-23 \pm \sqrt{(23)^2 - (4 \times 12 \times 10)}}{2 \times 12}$$

$$\implies x = \frac{-23 \pm 7}{24}$$

7. $15x^2 - 6x = 9$

$$\implies x = \frac{--6 \pm \sqrt{(-6)^2 - (4 \times 15 \times -9)}}{2 \times 15}$$

$$\implies x = \frac{6 \pm 24}{30}$$

8. $4x^2 - 17x = 15$

$$\implies x = \frac{--17 \pm \sqrt{(-17)^2 - (4 \times 4 \times -15)}}{2 \times 4}$$

$$\implies x = \frac{17 \pm 23}{8}$$

9. $8x^2 - 19x - 15 = 0$

$$\implies x = \frac{--19 \pm \sqrt{(-19)^2 - (4 \times 8 \times -15)}}{2 \times 8}$$

$$\implies x = \frac{19 \pm 29}{16}$$

10. $7(x + 2a)^2 + 3a^2 = 5a(7x + 23a)$

$$\implies 7x^2 + 28ax + 28a^2 + 3a^2 = 35ax + 115a^2$$

$$\implies 7x^2 - 7ax - 84a^2 = 0$$

$$\implies x = \frac{--7a \pm \sqrt{(-7a)^2 - (4 \times 7 \times -84a^2)}}{2 \times 7}$$

$$\implies x = \frac{7a \pm 49a}{14}$$

Question 14.3: Solve the following equations:

1. $x^2 + 2x - 3 = 0$

$$\implies x = \frac{--2 \pm \sqrt{2^2 - (4 \times 1 \times -3)}}{2 \times 1}$$

$$\implies x = \frac{2 \pm 4}{2}$$

2. $x^2 - 2x - 1 = 0$

$$\implies x = \frac{--2 \pm \sqrt{(-2)^2 - (4 \times 1 \times -1)}}{2 \times 1}$$

$$\implies x = \frac{2 \pm 2\sqrt{2}}{2}$$

3. $x^2 - 3x - 5 = 0$

$$\implies x = \frac{--3 \pm \sqrt{(-3)^2 - (4 \times 1 \times -5)}}{2 \times 1}$$

$$\implies x = \frac{3 \pm \sqrt{29}}{2}$$

4. $3x^2 - 2x - 1 = 0$

$$\implies x = \frac{--2 \pm \sqrt{(-2)^2 - (4 \times 3 \times -1)}}{2 \times 3}$$

$$\implies x = \frac{2 \pm 4}{6}$$

5. $2x^2 - 9x - 4 = 0$

$$\implies x = \frac{--9 \pm \sqrt{(-9)^2 - (4 \times 2 \times -4)}}{2 \times 2}$$

$$\implies x = \frac{9 \pm \sqrt{113}}{4}$$

6. $3x^2 + 7x - 6 = 0$

$$\implies x = \frac{-7 \pm \sqrt{7^2 - (4 \times 3 \times -6)}}{2 \times 3}$$

$$\implies x = \frac{-7 \pm 11}{6}$$

7. $4x^2 - 3x - 14 = 0$

$$\implies x = \frac{--3 \pm \sqrt{(-3)^2 - (4 \times 4 \times -14)}}{2 \times 4}$$

$$\implies x = \frac{3 \pm \sqrt{233}}{8}$$

8. $6x^2 - 7x - 3 = 0$

$$\implies x = \frac{--7 \pm \sqrt{(-7)^2 - (4 \times 6 \times -3)}}{2 \times 6}$$

$$\implies x = \frac{7 \pm 11}{12}$$

9. $12x^2 - 23x + 10 = 0$

$$\implies x = \frac{--23 \pm \sqrt{(-23)^2 - (4 \times 12 \times 10)}}{2 \times 10}$$

$$\implies x = \frac{23 \pm 7}{20}$$

10. $x^2 - 9x - 90 = 0$

$$\implies x = \frac{--9 \pm \sqrt{(-9)^2 - (4 \times 1 \times -90)}}{2 \times 1}$$

$$\implies x = \frac{9 \pm 21}{2}$$

Solutions to Chapter 15

Question 15.1: If $3(4x - 5y) = 2x - y$, find the ratio of $x : y$.

$3(4x - 5y) = 2x - y$

$\implies 12x - 2x = -y + 15y$

$\implies 10x = 14y$

$\implies x : y = 7 : 5$

Question 15.2: If $\dfrac{5a + 3b}{4a + 5b} = \dfrac{2}{3}$, find the ratio of $a : b$.

$\dfrac{5a + 3b}{4a + 5b} = \dfrac{2}{3}$

$\implies 3(5a + 3b) = 2(4a + 5b)$

$\implies 15a - 8a = 10b - 9b$

$\implies 7a = b$

$\implies a : b = 1 : 7$

Question 15.3: Two numbers are in the ratio of $5 : 7$. If 9 is added to each of them, the resulting numbers are in the ratio of $4 : 5$. What are the number?

$a : b = 5 : 7 \implies a = \dfrac{5b}{7}$

$a + 9 : b + 9 = 4 : 5 \implies (\dfrac{5b}{7} + 9) : (b + 9) = 4 : 5$

$\implies 5b + 63 : 7b + 63 = 4 : 5$

$\implies 25b + 315 = 28b + 252$

$\implies 3b = 63 \implies b = 21$

$\implies a = 15$

Therefore, $a = 15$ and $b = 21$

Question 15.4: If $\dfrac{a}{b} = \dfrac{7}{6}$, find the value of $(3a + 5b) : (7b - 5a)$.

$\dfrac{a}{b} = \dfrac{7}{6}$

$\implies a = \dfrac{7b}{6}$

$3a + 5b = \dfrac{21b}{6} + 5b = \dfrac{17b}{2}$

$7b - 5a = 7b - \dfrac{35b}{6} = \dfrac{7b}{6}$

$(3a + 5b) : (7b - 5a) = \dfrac{17b}{2} : \dfrac{7b}{6} = 51 : 7$

Question 15.5: If $\dfrac{2x}{3y} = \dfrac{5}{4}$, find the value of $(8x - 7y) : (y + 8x)$

133

$$\frac{2x}{3y} = \frac{5}{4} \implies x = \frac{15y}{8}$$

$$8x - 7y = \frac{120y}{8} - 7y = 8y$$

$$y + 8x = y + \frac{120y}{8} = 15y$$

Therefore, $(8x - 7y) : (y + 8x) = 8 : 15$

Question 15.6: If $16a = 25b$, find the duplicate ratio of $a : b$.

Duplicate ratio of $a : b = a^2 : b^2$

Given $16a : 25b \implies a = \frac{25b}{16}$

Therefore, $a^2 : b^2 = \frac{625b^2}{256} : b^2 = 625 : 256$

Question 15.7: If $25x = 9y$, find the subduplicate ratio of $x : y$.

$25x = 9y \implies x : y = 9 : 25$

$\implies \sqrt{x} : \sqrt{y} = 3 : 5$

Question 15.8: If $\dfrac{x}{cm - bn} = \dfrac{y}{cl - an} = \dfrac{z}{bl - am}$, show that $ax - by + cz = 0$

Let $\dfrac{x}{cm - bn} = \dfrac{y}{cl - an} = \dfrac{z}{bl - am} = k$

Then: $x = k(cm - bn), y = k(cl - an), z = k(bl - am)$

Substituting: $ax - by + cz = ak(cm - bn) - bk(cl - an) + ck(bl - am)$
$= k(acm - abn - bcl + abn + bcl - acm) = 0$

Question 15.9: Solve the following equations:

1. $(3x - 5) : (5x - 11) = 2 : 3$
   $3(3x - 5) = 2(5x - 11)$
   $9x - 15 = 10x - 22$
   $x = 7$

2. $(2x + 1) : (x + 5) = (6x - 7) : (3x + 5)$
   $(2x + 1)(3x + 5) = (6x - 7)(x + 5)$
   $6x^2 + 13x + 5 = 6x^2 + 23x - 35$
   $10x = 40$
   $x = 4$

3. $(3x - 2) : (x + 2) = (5x - 2) : (x + 8)$
$(3x - 2)(x + 8) = (5x - 2)(x + 2)$
$3x^2 + 22x - 16 = 5x^2 + 8x - 4$
$2x^2 - 14x + 12 = 0$
$x^2 - 7x + 6 = 0$
$(x - 6)(x - 1) = 0$
Therefore, $x = 1$, $x = 6$ are the solutions.

4. $x : y = 3 : 4 = (x + y) : (3x + 1)$
$x = \dfrac{3y}{4}$
$3(3x + 1) = 4(x + y)$
$9x + 3 - 4x - 4y = 0$
$9x + 3 - 4x - \dfrac{16x}{3} = 0$
$27x + 9 - 12x - 16x = 0$
$x = 9$

Note: For all these problems, we will use the following facts
$a : b : c \implies a : b = b : c = k$, therefore $a = bk = ck^2$

Question 15.10: If a, b, c be proportionals. show that:

1. $a - b : b - c = b : c$
$\dfrac{a - b}{b - c} = \dfrac{ck^2 - ck}{ck - c} = \dfrac{ck(k - 1)}{c(k - 1)} = k$
$b : c = k$
Therefore, $a - b : b - c = b : c$

2. $a + b : a - b = b + c : b - c$
$a + b : a - b = ck^2 + ck : ck^2 - ck = k + 1 : k - 1$
$b + c : b - c = ck + c : ck - c = k + 1 : k - 1$
Therefore $a + b : a - b = b + c : b - c = k + 1 : k - 1$.

3. $a + b : b + c = a : b$
$a + b : b + c = ck^2 + ck : ck + c = ck(k + 1) : c(k + 1) = k$
$a : b = k$
Therefore, $a + b : b + c = a : b = k$

4. $ma + nb : mb + nc = ma - nb : mb - nc$
$ma + nb : mb + nc = mck^2 + nck : mck + nc = ck(mk + n) :$
$c(mk + n) = k$

$ma - nb : mb - nc = mck^2 - nck : mck - nc = ck(mk - n) :$
$c(mk - n) = k$
Therefore, $ma + nb : mb + nc = ma - nb : mb - nc = k$

5. $a^2 + b^2 : (a + b)^2 = b^2 + c^2 : (b + c)^2$
$a^2 + b^2 : (a + b)^2 = c^2 k^4 + c^2 k^2 : (ck^2 + ck)^2$
$= c^2 k^2 (k^2 + 1) : c^2 k^2 (k + 1)^2 = (k^2 + 1) : (k + 1)^2$
$b^2 + c^2 : (b + c)^2 = c^2 k^2 + c^2 : (ck + c)^2$
$= c^2 (k^2 + 1) : c^2 (k + 1)^2 = (k^2 + 1) : (k + 1)^2$
Therefore, $a^2 + b^2 : (a+b)^2 = b^2 + c^2 : (b+c)^2 = (k^2 + 1) : (k+1)^2$

www.ingramcontent.com/pod-product-compliance
Lightning Source LLC
Chambersburg PA
CBHW051532170526
45165CB00002B/702